Roman Dialogues
by
Michael W Hurley

CHAPTER 1

Creedo woke up excited about the upcoming day. It was his day to make the speech. He ate a quick breakfast served by his favorite slave; and hurried to meet up with his colleague Vitellus. Creedo was from a proud patrician family; growing wine as far back as the divine Augustus. Creedo was about to give the main speech. The topic was the all important subject of water rights from the main aqueduct in Rome. Creedo would argue that farmers such as himself have as much rights to the aqueduct's water as the citizens and bathhouses of the city. He and his friend were walking towards Capitoline Hill--where the senate gathered--while musing about the day's business.

"Looking forward to your speech Creedo?" Vitellus remarked.

"Of course, I have prepared almost a whole week for this. Water rights is very important to me. The decision will determine the future of agriculture in Rome. Whether small farms like mine will survive or die. We must have equal access to the water from the aqueducts. I am looking forward to giving the speech I have prepared." Creedo responded.

"I hate to change the subject Creedo but I had a strange dream last night. I dreamt you'd become part of an empire as big as Rome. Not as a traitor of course, but, well, how shall I say this many years in the future," Vitellus revealed.

"That's nonsense Vitellus, there will never be an empire as great as Rome even if I live a thousand years. And neither you or I shall see anything a thousand years from now," Creedo proclaimed, eager to change the subject.

"Perhaps it was just a dream, not a prophecy. Do you think there will ever be an empire as great as Rome. Even in say two thousand years?" Vitellus asked, clearly not taking the hint.

"If I could see two thousand years from now and live in an empire nearly as great as Rome; then by Jupiter and Minerva may I see it! Till then let's just discuss the business at hand and hope we have the votes in the Senate to win on this issue!" Creedo demanded.

"Sure Creedo. I am looking forward to second your motion on the floor and persuade the other senators."

"Thank you Vitellus. You've always been a good friend. I can always count on you."

"Same to you Creedo. Same to you!"

CHAPTER 2

We worked on this infernal machine for several years now. A time displacement unit or "time machine" as it's commonly referred to; it was finally completed by our team of scientists at our MIT lab in Boston. I dedicated my life to temporal physics, and was determined to create a mechanism to transport any object through the past or future. We received funding through the Department of Energy as well as private donations. I was not allowed to look too closely at the motives behind our sponsors, but I personally suspected some military connections. When the unit finally became operational I decided to be the guinea pig. On the decision for time and place for my destination--Roman history being a personal interest and hobby of mine--I decided to go back to the Roman empire. My team approved of that so off I went. Needing to test my time machine prototype I had to transport myself to the past; and transport a person from the past to the present time. Only this would ensure the device we built was operational.

Creedo was speaking to a group of about 200 senators on Capitoline Hill. He was a good speaker from what I could tell, though my Latin was a little rusty. He had an an ambitious voice, but not angry or frightening. Deep baritone that resonated throughout the chamber. He was wearing a white, flowing tunic, or a tunica clava as they called it with a purple band around his shoulder. He was wearing a silver ring with the familiar initials SPQR on it for Senatus Populusque Romanus (the senate and people of Rome). I set up a hidden camera inside as well as an audio sensor to view and hear the debates about 300 yards from the Curia Hostilia, the place where the Roman Senate convened. I studied a great deal of Latin before I made this trip. I wanted to find the right person to take back with me to my small New Hampshire town in 1991.

I waited outside nervously, quivering back and forth. I hoped I would not have to use weapons to capture him but I came prepared. My machine had the best equipment money could buy. RPG's, 50 Calibur fire arms, laser guided rifles, and other weapons that could fend off a legion. I was holed up in this contraption, suffocating from impatience and excitement. I was in a huge time machine. More the shape and size of a fuselage of a 747 than a DeLorean or phone booth like those silly 80's movies, this thing needed to coordinate all the power of the sun to harness the fusion to demolecularize me through time. So I waited to pounce like a tiger. He came out in his fancy toga strutting out of the chamber. A satisfied grin on his face, they just passed a resolution he sponsored. The last thing he expected was a man from about 2000 years in the future sneaking up behind him whisking him away to another time and place. I came up behind him stalking my prey. An old cotton rag with ingrained grease stains reeked of chloroform, I shoved it up his nostrils with my right hand. He was coming with me, oh yes he was coming with me.

As an MIT professor in quantum physics I'm also an avid watcher of the History channel with a great deal of interest in politics and the similarities between civilizations. I was watching several programs on the Roman empire and the similarites between their society and ours. So I took it upon myself to create a time machine to go back to the height of the Roman imperium (around the 2nd century AD), transport a Roman citizen; preferably a senator or an educated aristocrat, and discuss with him the political, social, military, and economic parallels between Rome and the United States of America.

I taught him English obviously. At my university I presented him to some professors who taught Latin and had him learn the English language. I don't need to tell you how amazed they were to meet someone who actually spoke the dead language, who lived in Rome and spoke and dined with emperors. After a while he learned our language quite well, taught Latin in a college and blended in like any other American. I had him watch television, read papers, and explained American history from the time of Columbus to the present.

It was the time of emperor Traijan. A worthy emperor he reigned under the pax Romana at the height of the Roman empire. He built a new Roman forum (forum Romanus), a large marketplace, one could argue the world's first shopping mall, and was generally considered a decent and competent leader. This is the era that Creedo lived, worked, and was familiar with. I was dying to pick his brain and see what insights he had not only on Roman society, but our own.

Observing our country, its government and leadership, his opinions and observations lived up to my greatest expectations.   After the english lessons at MIT and Harvard as well as some grammer schools this is the first dialogue I had with the senator. This was a conversation I had over dinner in a small bistro in my small New Hampshire town of Littleton called Mise en Paris sometime in 2003.

"So what do you think about America so far?" I asked curiously.

"It is Rome transplanted 1500 years more or less," he shot back.

"In what ways?" I excitedly queried.

"How can you even ask that. It's obvious. You have a senate, many amphitheaters, a Caesars Palace, a plutocratic government, bread and circuses for the masses, and an emperor."

"Well we don't really have an emperor," I replied defensively..

"You chose a leader who was the son of a previous leader, chosen by his brother's region, unelected by the people, who invaded countries for no reason but power and glory; what do you call that dumbass?"

"Where did you learn that word," more puzzled than amused I replied.

"My students at school. They taught me lots of fun words like that shall I share them with you?" he said amused with himself.

"No that's okay, I know them all and am sorry you had to pick them up I wish you would forget the more profane words in the English language." I answered.

"Why I find them the most honest words in your language. The most real," he stated.

"Maybe so but try not to repeat them with me. I'm writing a literary piece and I'm not writing for people who appreciate such language." I said almost apologetically

"As you wish," Creedo replied.

"So as one who lived in a society where people had to fight to the death for people's pleasure with swords and maces and other such brutal instruments how can you justify that aspect of Roman culture?"

"No different than your boxing or wrestling," the senator answered.

"But people don't get killed in boxing or wrestling," I said.

"That's not true at all, people get killed all the time in boxing and wrestling. I read it in your papers and the box you people call television," he said.

"But they're not supposed to get killed in a boxing or wrestling match," I shot back at him.

"They weren't supposed to get killed in gladitorial combat either, many matches all parties lived. We had more than just that we had mock naval battles, great shows, the Circus Maximus had great chariot races, we really had a sense of grandeur and awe, something you people lack," he stated.

"I know, I wish we would have chariot races like that today, have you seen the movie Ben-Hur?" I questioned.

"No who is that?" he asked.

"No one real, that's a movie starring Charleton Heston an actor from the 1950's," I replied.

"Is he the one who became president of your country?" he inquired.

"No that was Ronald Reagan, but close though, they were both friends and Republicans," I responded.

"We had an actor/leader also, Nero, didn't work out too well," Creedo said shaking his head.

"I know he wasn't a very good emperor. Now I know we've made progress; I live in a country that would never burn people alive for their religious beliefs at the hands of the government, the way Nero did to the poor Christains," I confidently stated.

"Waco," he pronounced. Referring to the ATF fiasco in Texas.

"Touche`, guess we haven't made much progress after all," I said softly.

"Technologically perhaps, but politically not at all," he answered.

"Your society had slavery how can you justify that?" I asked.

"How can you? Your country had slavery until 1865. And our slavery wasn't based on color like yours was. This was part of our economic system, it's how we built our great aqueducts, amphitheaters, bath houses and structures of our day with low unit labor costs. Slaves had food, clothing, and shelter much better off than your homeless people I'd say," Creedo replied petulantly.

"Perhaps you're right, but I can't justify slavery," I muttered.

"We did what was best for us in the times when we lived, that's all any society can do," he stated.

"Good point." Was all I could say.

"I don't like your country's attitude about sex at all. Ours was more open and honest," Creedo proclaimed.

"How so?" I asked.

"We had legalized prostitution, orgies, toga parties, bath houses, and murals showing sexual images right in our homes," he informed. "Why is your society so prudish 1500 years after the Roman empire?" he asked inquisitively.

"I have no idea, good point." I stated.

"Of course human beings love sex. Why can't your religions, governments, and people just be honest and admit that? Why are so many of your Americans against this?" he questioned.

"Honestly, I don't know. Based on our Judeo-Christian puritan backround I suppose." I posited.

"Well get rid of it. Emperor Augustus tried that family values/moral majority crap early in the first century and we discarded it thankfully for a more open attitude towards carnal pleasures," he lectured.

"Well maybe someday, it's amazing that we seem to have gone backwards not forwards in that area. Watch the profanity please," I pleaded.

"Sorry, I'm just passionate on this subject," Creedo stated.

"That's okay, like you said these words convey honesty and I guess the word crap isn't that bad," I replied.

"So what is your country contributing to the countries it dominates. We gave our dominions a water system, amphitheaters, bath houses, culture, law and order. What is your nation doing in that regard?" He asked me piercing his eyes at me.

"I don't know to be honest with you democracy and freedom I suppose?" I hesitantly answered.

"Like Cuba?" he questioned.

"Well that was a hundred year old venture, I suppose the Spanish-American War didn't work out that well regarding Cuba," I reluctantly answered.

"And how will your adventure in Iraq go?" he asked me (this conversation took place soon after the war started).

"I don't know we'll see in a hundred years I guess," I said almost jokingly.

"I suggest just be honest about it like we were, declare yourselves an empire, invade countries, take their peoples slaves and plunder their wealth," he said seriously.

"We can't do that in modern day America!" I declared.

"You already live in a defacto empire, invading countries for no reason, your leader goes to war with little or no consent of your legislature, breaks laws with no consequences, just admit it and pillage and rampage till your hearts delight!" He then laughed, but was serious about the proposal.

"Maybe you have a point. You seem to have a theme that it's better to be honest about things than sugarcoat them with words such as making the world safe for democracy or spreading liberty in the Muslim world." I stated to the senator.

"I won't use profane words like you asked but I believe the answer to your country's presumption of making the world safe for democracy and spreading liberty in the Muslim world is answered by the initials B.S," he said bemused with himself.

"So what do you think of the recent attacks on our country? Does that remind you of the barbarians attacking ancient Rome?"

"This is the start of a long war between the civilized empire and the tribes of the muslim lands. This is a war that will last centuries--I swear by the gods this is true," Creedo proclaimed.

"Who do you think will win?" I asked.

"Whichever side realizes this," he answered.

"You're very funny I would have liked to have known you in the Roman empire and experienced it for myself. Maybe I'll take my time machine and live back then for a while, learn latin better and experience the glory and imperial majesty of that era," I said to Creedo.

"It was a great time I'll show you around. You'd love the bath houses and our wine was to die for. I'll show you the games as well," he said wistfully.

"I don't think so I'm squeamish." I said sincerely.

"You know I'm quite certain if your people had real gladitorial combat, people being eaten by lions, and live executions, it would be sold out in minutes," he stated to me.

"You're probably right our movies and television are just as violent. Graphic blood, people would probably pay to see real blood even today," I philosophized.

"People don't change, they're either honest about what they are violent, sexual beasts, seeking pleasure, or not. All we said it's better to be honest about it and go with it," he stated.

"You really have a point, but executing people in arenas isn't right," I said.

"You still have the death penalty today why not have fun with it?" Creedo proposed.

"I suppose you make another good point. The more things change the more they stay the same," I replied back.

"They haven't changed. Like I said we were honest about what we were and you are not," he said to me.

"You are a very enlightening figure. I'm afraid it's 11 and I have to go to bed. Let's have lunch tomorrow ok," I asked hopefully.

"Ok but not at that McDoogles place, honestly I prefer roasted mice and snails," he answered mispronouncing McDonalds.

"So would most Americans," I responded.

CHAPTER 3

We had lunch the next day. But not at McDonald's or as he said McDoogles but at a fine bistro in town owned by a local chef whose food is far superior to any fast food chain. I thought all night about our conversation and realized that American presidents over the last hundred years acted exactly as Roman emperors did 1500 years ago. Not for democracy and freedom as they claim but for hegemony, imperialism, and domination. It was a lot of food for thought and I couldn't wait to have another discussion with him.

"I have a question for you. The United States started off as a colony, and so did Britain under your empire. Why do they then become empires of their own rather than adopt a Swiss like noninterventionist policy towards other nations?" I asked him.

"Some countries seek independence so they can dominate other states of their own. Also they practice what they've learned from the colonizer and do that to other countries. Brittania was our colony for hundreds of years. They studied our empire thoroughly and in the 18th and 19th centuries established a great empire of their own tracing its roots unquestionably to the Roman empire." he replied.

I stretched my arms and stated,"Now what about the Nazis were they trying to recreated the Roman empire in their own twisted way?"

He pointedly stated, "That wasn't comparable to Rome at all. Just a failed art student with delusions of grandeur. The Huns could never do what we did. Their empire lasted about 4 years and fortunately they were crushed. The only state that could do what we did for 500 years is the United States if you guys get your act together."

"How so?" I queried.

"Well get your debts under control. And when you conquer a territory keep it. From what I read it's better to keep a territory than let it become independent, from the history books I've read," Creedo stated.

"Really?" I whimpered.

"Sure. Let me list all the countries that the US invaded or overthrew a government; Hawaii, Cuba, Puerto Rico, Nicaraugua in 1910 and the 1980's, Phillipines, Guatamala, Iran in 1954, Panama (twice), Chile, Grenada, and Iraq. The only places of those listed that are currently prosperous, stable, and viable are Hawaii and Puerto Rico right?" he propositioned.

"That's true," I mustered.

"And that's because they're a state or US territory. They get all the benefits from becoming part of a great empire; when you leave a place on its own after invading it you just destabilize it and make them resent you. That's why countries like Iran and Cuba still hate the US. If you annex the territory and make it part of your empire they get the benefits of being integrated into your country which is better off for everyone in the long run," he proclaimed.

I sat up in my chair to ask,"So should Iraq and Afghanistan become part of the US?"

"Absolutely and they should be the ones demanding it. These are failed states that would reap the benefit of such an action." Creedo answered sipping a latte.

"Interesting. I know Puerto Rico gets lots of tax advantages from being a territory. Tariff and trade advantages as well. Also they can come to the mainland with no visas or passports. Maybe you have a point." I said.

"Of course I do. Everytime the US causes a regime change they should annex the territory and make it part of the US as a territory or state. We knew that 2000 years ago," he lectured.

"Well this was another great discussion Creedo I have a lot to think about. How was your meal?" I wondered.

"Very good beni, I love Italian food," he had a satisfied grin on his pasta covered face.

That evening I had much to think about. I started thinking about the obvious similarities between the Roman and American empires. Both started off as small agrarian states that became vast industrial and military empires with worldwide influence and power. I wondered if our founding fathers intended a little colonial outpost to become the most powerful and influential country in the world. It certainly would have seemed impossible at the time. To imagine the 13 colonies becoming more powerful than Europe would've seemed ridiculous and laughable. Europe was the undisputed center of the world with most of the world under its control. It would've been like saying sub-Saharan Africa will be the most powerful region on earth or Australia. But it happened. The United States grew just like Rome did. Westward expansion through wars, treaties, and displacement of native peoples quadrupled the size of the US. While Europe fought among themselves with the Napoleonic wars, WWI, WW II, and other conflicts, the US was quickly using their divisions to amass more power. Except for the Civil War the continental US was always pretty much unified while Europe was and still is not. Recently Europe tried to be unified but so far it seems to be mostly paper treaties and nominal alliances. I don't see the European Union being a threat to American hegemony and I think this is not because I love the United States but because of the difficulty of unifying a region as diverse and with such national, ethnic, and language differences as Europe. China is quickly developing into an economic powerhouse. They may be America's greatest competition in the near future.

Creedo seemed quite the imperialist. Being a senator in ancient Rome I can see why. He was born to wealthy aristocrats on a large farm near Rome. They had many slaves but the family treated them well. They sold wine to the local populace. It was good wine, very sweet and got you intoxicated very quickly. Wine and baths were in vogue back then and that's what Creedo liked to do for fun. I really plan to use my time machine and stay in ancient Rome for a while. I have to learn latin first, it's a difficult language and it'll take a while to learn but I have to know it to fit in with the ancient Romans and hopefully participate in their daily activities. I plan to do this after showing Creedo off to certain people. People who could appreciate what he knows and apply ancient Roman thoughts to modern day policies. I also plan to get more people from ancient Rome perhaps experts in architecture, engineering, medicine, art and other areas and have them talk to present day experts for a greater comparison of these studies than what we found from artifacts from that era. I musn't forget that most of the knowlege from that era was lost from the great fires that burned the Library of Alexandria where most of the great works from Rome, Greece and others were lost forever. But NO MORE they will be recovered. I will carry a team of experts to retrieve these great works and they will be studied profusely and we will know more about the ancient world than ever before!

I introduced Creedo to a history professor at a local college nearby. I wondered if he had any more insights than I had or could ask better questions. I was present in the room at the time so I could record his answers with a pocket tape recorder. Here is an except of that conversation. The professor's name was Harold Brown who was renowned for Roman and ancient history.

"It's an honor to meet you Creedo would you like some coffee or tea?" the professor declared.

"Tea would be fine," said the senator.

The professor started off. "First of all what was life like in ancient Rome?"

"It was wonderful if you were a patrician, if you were a slave or a plebe it wasn't so great. Things haven't changed much in that respect." stated Creedo.

"That's certainly true. The more things change the more they stay the same. What did you do for pleasure, recreation?" asked Brown.

"Watched the games, gladiator matches, chariot races, baths, orgies, wine or to you people boxing, NASCAR, and sex, drugs, and rock and roll!" yelled the Roman.

Professor Brown replied "Well I don't partake in such pursuits but the analogy is clear our passions and pastimes haven't changed much. Now what do you think of our present government as compared to ancient Rome?"

"I think your government is trying actively to be the next Rome. Trying to achieve worldwide hegemony as you call it and I advocated doing it more directly and honestly than trying to sugarcoat it with terms such as trying to spread democracy or liberating countries," Creedo surmised.

"What do you think we should do then?" asked the historian.

"Expand your country. Start by annexing Canada, then Mexico. That would solve your illegal immigration problem. The Mexicans wouldn't have to move to the US if they were already here. And your country could share it's resources and wealth with Mexico if it became part of the USA. Your government could also annex central and south America and call it the United States of the Americas. A united hemisphere would be a powerful entity don't you think?" said Creedo grinning.

"You're quite the imperialist Creedo. That would solve the immigration problem though. People wouldn't need to come to the United States if they were already here. We would be responsible for a lot of poor people in other countries though." Professor Brown came back with.

"Yes but they would stay there. The government could send a team of experts to help these countries, rebuild infrastructures and then use the cheap labor to benefit the empire just as we did," shot back Creedo.

"I hope you don't talk to our current leaders. They just might go for a proposal like that," worried Brown.

"I would very much like to talk to your president and leaders. I think they would have a lot to learn from me and ancient Roman government," said Creedo pacing the floor.

"Now you lived in the first century AD right?" asked Brown.

"According to your calendar that's right," he replied back.

"I assume you read about what happens next? The fall of the Roman empire in the 5th century and the sacking by the Vandals and the Visigoths." informed the professor.

"The Roman empire never died. Like matter it only changed in form. It spread to Constantinople and the Byzantine empire. The center of the Catholic church with hundreds of millions of followers is still there, it's the capital of Italy and the spirit of Rome spread to the British, Spanish and American empires years later. Rome will never die, it's the eternal city," Creedo proclaimed.

"Good answer and very true. Do you think America will have the same impact 2000 years from now?" asked Dr. Brown.

"If you follow these rules and become more honest and forthright about becoming an empire and conquer more territory," Creedo mused.

The professor nodded, then said. "Empires get a bad rap today. But you're saying they do more good than bad."

"Absolutely. When the Roman empire officially if not really collapsed in the 5th century, Europe entered the Dark Ages. Feudalism tried to replace the order of the Roman empire but never quite succeeded. Technologically, culturally, and politically Europe went backwards for a thousand years because of the collapse of the Roman empire. So what I want America to do is formally conquer as many states and nations as possible to recreate the Roman empire in all its glory which will actually be more beneficial to the lands conquered than to American itself." Creedo said dryly.

"It's scary but I'm actually listening to this." said the professor.

"I have read some great modern books that agree with me on this subject." Creedo stated.

"Like what Mein Kampf" said the professor.

Creedo shrugged and said "No that stupid hun could never have succeeded. It's amazing he got as far as he got. Napoleon coudn't have succeeded for long either. Only the United States. Did you know that the US military budget exceeds that of the rest of the world combined. And all they did recently is invade Afghanistan and Iraq. That would be like having a million dollars and buying a trailerhome and a beat up pickup truck. You should conquer much more territory with that kind of money."

The professor stated back "Well the United States should be more defensive than offensive in our foreign policy in my opinion. Hence the term Defense Department."

Creedo shot back "That in your word is bull. It's hardly defensive."

"Yes you're probably right. Look I find you fascinating. I have a friend in the state department who's a Roman history buff who would love to chat with you. I will give your friend his number and contact information so you can talk to him sometime." The professor proposed to the senator.

I then interjected and said "I'll get on this right away. Creedo would have a lot to say to those in government."

Creedo stated firmly, "I absolutely do."

Chapter 4

She was in the second row of the classroom. Not a great student but not a poor one either. Flaxen hair, moist skin, lightly tanned, sea blue eyes and wearing a cleaveage baring blouse and skintight jeans. Mary Burns was her name as if that mattered. Creedo was teaching Virgil's Aenead in the original Latin. She had some obscure question regarding conjugating a verb. But the gleam in her eyes said something else. After class was over she wanted to see him in his office. Creedo was a good looking man. Copper skin, stocky build he had the look of a powerful yet sophisticated aristocrat. Dark hair and penetrating eyes he had the look women could not resist. Women still love Italian men even today and she would devour the chance to be one on one with him.

She entered the room and asked a simple question she knew the answer to. She then flipped her hair in a way that said take me. Creedo got the hint. Soon they were making love on his couch. He ripped her blouse off and conquered her only a Roman could. She screamed in ecstacy as she was being ravished. At the end he proclaimed "Now you've had the original Latin lover." She wouldn't worry about her marks in that class again.

Creedo liked teaching. He liked interacting with young Americans. He was 52 but thought and acted younger. He thought of himself as the perfect Latin teacher and the students thought so too. The only people he could really talk to were people high up in the Catholic church and Latin teachers in his original tongue. He learned english quick but still preferred the dead language of the Caesars. After a few months though he felt a higher calling. He felt that I chose him to travel in my time machine for a greater purpose. A purpose as great as Rome itself. He told me his goal is to somehow create the world's most powerful empire and have it assert itself for a very long period. I just thought he was talking ancient history at first, but I later found out what is to come.

Chapter 5

Charles Taylor Huffington is a stout man in his late 40's. A career bureaucrat, went to Yale for international relations and started work for the US Ambassador in Paris. After that he transferred to the state department's national securtiy division and rose up to Undersecretary for National Securtiy where he is today. I called him to tell him about my time machine and that I had a Roman senator who wanted to speak with him. Of course he didn't believe me. But after a while I explained how I built the time machine by accessing transdimensional wormholes inside all time-space fabric and he agreed to meet with Creedo and have a discussion. I brought hime to Mr. Huffington's office where we had an appointment. The conversation went like this.

Creedo stepped in the office briskly "Well how do you do?"

"Very good. I hear you're a roman citizen. Not a modern one but an ancient one from a long time ago." the bureaucrat spoke.

"Absolutely. And I have things I want to tell you and your government." Creedo pointedly remarked.

"And what is that." Huffington replied.

"Honesty sir. Just be honest about what your country is." spoketh the senator.

"What do you mean?" said Mr. Huffington puzzled.

"I mean your country is an empire and your government should admit that. Ever since the beginning it expanded through manifest destiny and like Rome expanded ever since. The territories you annexed like Texas, California, Hawaii, etcetera have been far better off than the territories it invaded and left like Cuba, Phillipines and several Latin American countries." said Creedo.

"I'm impressed, you must've studied a hell of a lot of American history since you been here." Surmised the official.

Creedo stated "Constantly. I see the parallels between American and Roman society and have some suggestions that I want to give to your government for the betterment of your country and the world."

"Well I'm glad to receive any advice from someone as knowlegeable as you. Go on." he said to the senator.

"Okay, I could go on and on about this but basically I think the United States should form an official empire and start by annexing Canada and Mexico." proposed the imperialist.

"Interesting. Why should we do that?" shot back Huffington tersely.

Creedo stood straight and waved his arms "For their sake really. Economic, military reasons. I have this idea. Form something called the American Union. You could invite Canada and Mexico to join. They'd initially think it's some kind of expanded version of NAFTA. But it would really be an expansion of the American state. All military, economic, political, and foreign policy decisions for North America would be made in Washington."

The undersecretary cleverly asked "What would be the difference between that and now?"

"It would formalize the relationship. This would allow Mexicans to be American citizens so they wouldn't migrate north. Canada, rich in oil and other resources, would be a great benefit to the United States. They would share their natural resources in exchange for military protection and we would set up large military bases on the northern US border and southern Alaska to ensure their accension to this proposal." Creedo suggested.

"Wow what next?" shouted Huffington.

"The American Union would acquire Latin and South America creating an American superstate consisting of all the Americas. One hemisphere under the control of the US through the American Union. This wouldn't be like he British commonwealth or the UN. This would be a real extension of the American state with the power to tax, control resources, foreign policy, trade and negotiate treaties in exchange for military protection and economic assistance. They would be a territory like Guam or Puerto Rico and after 10 years in the American Union they could have the option to become a state." stated Creedo's voluminous manifesto.

"You thought a lot about this. Why?" questioned Mr.H

"I want to look after my new country and help it grow like a flower," the Roman answered.

"You know our administration might like that idea. American Union sounds so innocuous. We could force countries to join and get their people to pay off our national debt and our foreign debts. Especially to China," the Washington official was clearly open to the idea and responded positively.

"Absolutely. Think of the revenue flow from all those people. Africa would certainly join. Just offer say 50 billiion in aid and end farm subsidies and they'd join in a heartbeat. They'd be much better off being run by the US," Creedo explained.

"But would we really want Africa?" queried Huffington.

"Absolutely. You could set up military bases very close to the middle east keeping a close watch on your oil supplies there. Africa is rich in gold, diamonds, oil and other resources and the US could exploit that as well as dirt cheap labor." explained Creedo.

"True enough. You have some really good ideas you should come work for us. I think I'm going to have my personal assistant type up a memo on this subject. I will give you credit of course." sounded Huffington impressed with the proposals.

I jumped in "I can't believe you're taking this seriously."

"Oh yes. We've been thinking about something like this for a while now. Just waiting for the right time to act. With Creedo here we could say this is a sign from God for such an expansion policy. To create a historic link between America and ancient Rome! I think it's fantastic," proclaimed the official smiling.

"Me too," said the senator.

I ended, "Well that makes two of you."

Chapter 6

"In the councils of government, we must guard against the unwarranted influence whether sought or unsought, by the military-industrial complex. The potential for the disastrous rise of misplaced power exists and will persist."
President Eisenhower January 17, 1961

About a year ago I took Creedo to Rome so he could compare and contrast modern Rome with the ancient world. I took him to the forum and the Coliseum. He also viewed the Vatican and other sites. He thought they should rebuild the forum and Coliseum as it was. He couldn't understand why they let such buildings decay. He said if there was a huge hole in the capitol building or the Statue of Liberty we would repair it. That was his suggestion as to what they should do in Rome as well as all ancient ruins. Rebuilding things to their previous glory is better than letting things rot he said. I could see his point I must admit. I also gave him a tour of Washington, DC. Even though it was after 9/11 and security was tight, Mr. Huffington gave us a special pass that let us see the inner sanctums of the Capitol, White House, Pentagon and other agencies and government departments. He was amused by the romanesque architecture and theme of our nation's capital. He suggested we build an amphitheater and I showed him RFK stadium. He then suggested Americans would have fun watching terrorists and other criminals being eaten by lions and fight to the death in a recreated coliseum. As I watched the cable news programs and see the demagogury towards certain people I think certain elements would accept that unfortunately. Walking on the mall I asked Creedo why he wanted the US to officially conquer the world. He said that it was going to happen sooner or later and the sooner it happened the better off the world would be. He said it should happen now before any power becomes big enough to challenge American supremecy and that would cost more lives in the long run.

He said an official empire controlled by America would usher in a long reign of Pax Americana that would last hundreds of years and be good for the world. He then quoted Cicero "In men of the highest character and noblest genius there is to be found an insatiable desire for honor, command, power and glory." I said that our leader lack character and noble genius and deserve no more command, power and glory. We both had a laugh and got some dinner. The next day Mr. Huffington gave us a secret tour of the Pentagon. It was in the E ring and it was a huge room with supercomputers, huge trinitrons and men with headsets talking to people. I asked Mr. Huffington what this room was for and he directed me to the Director of Media Affairs Richard Sandon. A worn man in his late 50's, a Marlboro in his mouth, he looked like a career bureaucrat who works long hours for his government pay.

"Hi, so you're the director here." I said as we were introduced.

"Absolutely", the director replied.

"So you run the media affairs bureau for the Pentagon." I stated.

"Not exactly we run the media." He flatly stated.

"For the military?" I asked pointedly.

"No we run all the media CNN, Fox News, ABC, CBS, NBC, MSNBC. All of it." He said.

"No way really?" I couldn't believe my ears.

The director proclaimed "Oh yes we've been controlling the news and the flow of information for about 25 years now. Format, copy, editing, headlines and issues are all originated and approved here."

"Great, I wish we had a Pentagon," the senator interrupted.

"But that can't be. What about the first amendment and then why is some of the news so negative about the government and politicians?" I asked, sure this was not the case.

"Various presidents have given us waivers to the free press rights. And some news is negative to give the news credibility. Also it gives politicians, especially right wing politicians, something to rail against when they criticize the government, the war, and so on." Sandon explained to us.

"Has anyone found out about this?" I asked gasping for more information.

"Everyone in the media knows it. We've flooded the cable new with conservative commentators and crap about murderers and sex offenders. We were surprised and elated at how well that's worked." the Director lectured.

"So that's why the news is about child molestors, crimes and celebrities. I can't stand hearing about that crap. That's why I watch C-SPAN and 'The Jim Lehrer News Hour' ." I said assuredly.

"So you're the one." jokingly stated Sandon. The room erupts in laughter.

"Why did you guys make the news so much about crimes then?" I asked.

"Well we were hesitant initially. We thought some people might be smart enough to see this diversion. But say when you hear about some 6 year old girl from say Kansas being raped, molested and murdered what is your reaction?" He asked smartly.

I shouted out "What a horrible bastard. Kill the SOB!"

"Exactly! You dont' think what has this got to do with the 8.5 trillion dollar national debt, the 4 trillion dollar foreign debt, the war in Iraq, the price of gas, the outsourcing of American jobs, global warming, or anything elso remotely considered a real issue." he wisely explained to me.

"That's right. I never thought about it like that. I never knew this was a government operation though to dumb down the news." I sadly realized.

"Absolutely. It's going much better than anticipated. Soon all major news will be nothing but sex offender, criminals, celebrities, and sports/entertainment. Ratings are great and people are eating it up." Director Sandon explained.

Creedo interjected and shouted "This is basically a circus. Leaders distracting the masses. Wonderful! Good job men."

I stupidly asked "How do you know I won't expose this?"

Director Sandon calmly stated "We'll label you a registered sex offender. You don't want that stigma do you?"

"Hell no. I'll keep my mouth shut about this." I nervously withdrew.

"Yes YOU"D be on the news, haha." Sandon finished chuckling.

After this strange conversation at the Pentagon, Creedo and I went back to the mall and stared at the Washington Monument. He told me it reminded him of Traijan's column in the Roman forum. That monument commemorates the conquest of Dacia by Traijan and his generals and the entire structure is covered with reliefs that scroll up the tower telling a story in pictures. In contrast the Washington Monument in DC is blank, all white with nothing on it. He asked me why that was and the best thing I could come up with is that it is a blank canvass because the future of America hasn't been told yet and is so vast it could not be told on a structure 10 times the height of the Washington Monument. He said that was a good answer as we walked along the mall.

Chapter 7

Professor Brown met with me in his office. He seemed petulant, nervously stroking the eraser of a pencil, calling me in as if he had something urgent to talk about. His office was a small, off-white room with the most basic furniture, basic pc and printer, and textbooks on an oak shelf. He started off twitching his pencil.

"I want to talk about Creedo, what's the deal with that guy," he started off.

"I don't know he's got these crazy Roman ideas, he's harmless really." I naively replied.

"So he's been talking to some officials in government I hear," he said curtly.

"Oh yah, some hypothetical discussions and stuff like that." I said back.

"I don't think that's right. His opinions might interfere with our own time too much. It will affect history and perhaps the time space continuum so that people will disappear or something." He said all too seriously.

"That's only true in ' 80's movies with sitcom actors. Trust me we're perfectly safe. Remember I'm the quantum physics professor," I said with a smile.

"Ok. That's good. His ideas are pretty strange though, I mean America taking over the world and everything. C'mon that's ridiculous," he said earnestly.

"No more ridiculous than what's going on really. Is our government's foreign policy really that different than Creedo's or for that matter ancient Rome's?" I shot back.

"Well I still think it's dangerous if they listen to him too seriously. Look I hate to bring this up, but there was this student who used to go to my school, now she goes to Boston University where Creedo teaches Latin. Her name is Mary Burns and she's my girlfriend. The grapevine says she's been, well seeing him. I hope it's not true do you know anything of this?" he nervously asked me, trying to read my eyes with his piercing at me.

"I don't know anything about this, sorry," I lied but I really didn't want to get involved. Creedo told me about the affair and at the time I had no idea she was Professor Brown's girlfriend. I really didn't want to know or care about such things. The historical, political, cultural, and temporal significance of dealing with an ancient Roman citizen is what I was interested in. The last thing I wanted to deal with is love triangles.

"Well I don't care if he's Julius friggin Caesar himself, if that's true I'm kicking his ass!" He passionately proclaimed.

"I'm sure it's not true. With all due respect I'd like to move on from this subject." I respectfully pleaded.

"Ok but give him that message asap ok," he commanded.

"Yes I will," I said anxious to drop the subject.

"Good. Look I really hate to say this but I have a bad feeling about that guy. I just feel an ominous presence around him. A sense of foreboding. I don't know what it is but I somehow think he's dangerous. I don't to who or what, but I just get this sixth sense you know," he confided to me.

"Honestly, I don't know what you're talking about. He's one of the most interesting and insightful people I've had the pleasure of knowing. His knowlege of Rome and the US is astounding. And hey he sure learned english well and quick. He even knows slang and profanity," I replied earnestly.

"Yes well english is based on latin. Many english words derive from latin. So does spanish, french, german, and of course Italian. Rome occupied all those countries and hence the languages are latin based," the professor informed.

"Ok. Well I seriously think Creedo means no harm. I hope to assuage your fears about him," I said reassuringly.

"I'm sure its nothing. But if he's Mary's Latin lover I'm opening a can of wupass on him tell him that!" he exclaimed.

"Ok, ok... I will next time I see him. Hey have a good one," I said walking out of the room.

"You too," he quickly answered.

After that unfortunate conversation with Dr. Brown I had more pressing things to take care of. I wanted to rev up my time displacement unit and save works from the great Library of Alexandria as stated earlier. I gathered a team of historians to come with me to rescue some of the great lost works by authors such as Sophocles, Aeschylus, Euripides, and Plutarch. This would take a while. But time was not of the essence since it is malleable to me. I could come back and forth at will and would do so. Creedo would stay in America in the present time while I did this. He was a very popular man.

Everyone in government wanted to talk to him and he was glad to talk to them and give his historical perspectives to them. I only demanded they would not reveal him to the media. That would ruin everything, erase his relative anonymity, and make him a marked and public figure, instead of the wealth of knowlege and information that I knew he was and is. The media would only bring hysteria and I didn't want that at the expense of history. Fortunately the figures he conversed with were glad to comply. I also wanted my time machine kept secret so it wouldn't fall into the wrong hands. I only wish the future held that to be the case.

Chapter 8

The phone rang in my office. It was a hard, guttural voice on the other end. He said he was from the White House and had an appointment to talk to Creedo and asked if I'd come along. I jumped at the chance and Creedo did too. He had something specific he wanted to relay to the president and the powers that be. The next day Creedo and I flew down to Washington and entered the office of the White House Chief of Staff. A dark haired man of medium build, he looked extremely professional, perfectly dressed and manicured, very metrosexual, he sat us down on a nice leather couch in the office when Creedo immediately began to speak.

"My name in Gaius Creedo Vespanius. A Roman senator under Caesar Imperator Trajan." Creedo said dramatically.

"I'm John Balducci White House chief of staff and it's a supreme honor to meet you Creedo, my ancestors are from Rome." The official said.

"Beni, Very good. I've got something I want to inform you, your leader, and your country," Creedo announced.

"What is that Creedo?" he replied.

"First I must explain how my country became an empire," he stated.

"How did it?" The bureaucrat asked.

"Well actually it never officially became an empire at all. After decades of civil wars, fighting between Roman generals like Marius and Sulla, Caesar and Pompey, and the Triumverates when the Republic was split between different factions; as you are probably well aware Octavian finally brought peace to the Republic and her territories by defeating Marc Antony and his consort Cleopatra at the battle of Actium in 31 BC," the senator informed talking like a history professor.

"Ok I knew that then what?" Mr. Balducci said.

"Then Octavian assumed emergency powers from the senate. In 27 BC the senate named him Augustus but thats not the important thing for you people. The senate gave Caesar Augustus the powers of various offices like consul, who was basically like a modern day prime minister, praefect or commander in chief, tribune which was a powerful office who could veto other magistrates and execute anyone who interfered with his duties, and censor, who was chief of public morals and could remove senators, as well as senatus princeps or president of the senate. He was called many things; princeps, caesar, dominus, imperator but he was never called emperor and Rome was always officially a republic. But Augustus and the future caesars were very powerful," Creedo declared.

"So he more or less had the powers of an American president the way I see it,"

Balducci admitted.

"Wait I have more to say. After September 11 your country passed a law giving the president the powers to execute the war on terror any way he saw fit. Now recenty the US Congress passed the Military Commissions Act which gives the power to the president to declare anyone an enemy combantant sending them to a detention camp and according to your attorney general torturing them until death," Creedo explained passionately.

"Yes, we had to work hard to get all that through. So how does a republic become an empire?" The advisor asked emphatically.

"In one word security. If more than half of the people prefer security to freedom a republic dies and something else takes its place. All great leaders know this," Creedo said knowingly.

"So what's the point you're trying to make," said the chief of staff not trying to sound impatient.

"The point I'm trying to make is this. That America has become an empire. The republic is now dead. You have a defacto empire. Like Rome this is an unofficial empire with republican institutions. In fact by the very standards used to declare Rome an empire after the triumverates and the civil wars, the United States by that standard has an emperor and is ready for the next steps in its history," Creedo barked to the man in the desk. The chief of staff looked shocked, but pleased by this statement.

"So what do we do now?" he leaned forward and mustered.

They were austere men coming into the cavernous laboratory facility at the place where I kept my time displacement unit. I was tweaking the exhaust unit when these two men came in asking for me. They were from the NSA and wanted such technology for government use. I assumed this was a secret project within MIT's research department, located underground beneath the deserts of Arizona, but apparently the government was tipped off and sent some of their men to investigate. I told them I'd rather continue this project for educational purposes but I'd let them know if I changed my mind. They said the government could give me and my group millions and perhaps billions in funding for a time travel project involving the government, but again I declined. I thought of all the military and wartime applications the government would use my time machine for and I didn't want it falling into their hands. But they knew where it was now and I knew they'd be back. My worst fears were realized and I knew that someday they'd be back and use this machine that could do so much good for destructive purpsoses; just like flight or nuclear energy. So as quickly as I could I moved the facility, it almost didn't matter where. I knew that it was inevitable with this genie out of the bottle they'd be back and twist its peaceful and civilian applications for their destructive and warlike ends. Just a matter of time I thought. Just a matter of time.

Chapter 9

He swung his right fist like swinging an axe hitting Creedo square in the jaw. It was outside in the parking lot of Creedo's university building. He then swung his left fist at Creedo's nose making it ooze with blood that soaked his shirt and dazed his brain. Creedo got up and grabbed Professor Brown around the neck in a strangle hold. The professor struggling to get free just squirmed like a mouse caught in a trap. Creedo was bewildered by this. So still bleeding he shouted.

"Why are you doing this Dr. Brown," Creedo wondered.

"You know damn well why. Messing around with my girl. You leave her alone you son of a bitch," the professor exclaimed.

"I don't know what you're talking about. What girl?" Creedo hesitantly replied.

"Don't play dumb with me you've been messing around with Mary Burns my girlfriend. Trust me I have ways of finding out about this kinda stuff. You stay away from her you got it!" Dr. Brown forcefully stated.

"She doesn't love you, you know," Creedo mumbled almost to himself. "She loves me, you don't show her the time or affection."

"You better leave her alone or they'll be hell to pay. If you EVER see her again I'll kill you. I don't like you, I don't like your wacky ideas, and I never want to see your roman ass ever again you got that," Professor Brown screamed at the top of his lungs.

"Fine by me. I stopped seeing Mary a while ago. I don't think you're man enough for her though," Creedo shot back.

"Fuck you roman," Dr. Brown said.

"Et tu Americanus," the senator finished.

Creedo told me about this sad state of events the next day. I really didn't want to get involved with any love triangles, and having been friends with both of them I didn't want to take sides either. I just hoped that both being well educated, reasonable men, that cooler heads would prevail. There was obviously bad blood between them. I thought it was more than just Mary Burns and the cheating. Dr. Brown just felt a dark, ominous fear about Creedo that he couldn't put his finger on. Something that would change the destiny of him and perhaps something much greater. I felt Creedo's power too. But I thought it was a good thing. To enlighten people about the past, to learn more about the present, and positively affect the future. I thought his destiny would be benign. It certainly seemed so at the time. Only time would tell.

Creedo was licking his wounds. After the confrontation he felt languid and distressed. He thought of ways to cheer himself up. After some thought he had a moment of inspiration and said quietly to himself, "*Orgia"*.

He knew plenty of women from the college. Students, teachers, T.A's -no pun intended, sorority sisters, and some of Mary's friends. He invited them all to his den of lust. It would remind him of the good old days of Rome. Many students were eager to learn firsthand what a real roman orgy would be like. They were about to find out. He invited men too. The romans had no hangups about bisexuality. Many emporers had male lovers. Creedo was not personally into sex with men, but he invited all people to his orgia or orgy. Whatever they wanted to do was fine with him. By the time he was done with invitations almost a hundred people were willing to attend. He declared it a festival to Minerva and Dionysus the god of wine and sex. When they arrived to his off campus house they had to remove their shoes. He designated some people slaves who would wash their feet and lick the dirt off their naked bodies. He called the men satyrs and the women nymphs and set up an extravagent dinner. Contrary to popular belief roman orgies were more about food than sex. He served a seven course meal and went to specialty stores to get some unusual foods. They ate with their hands just like the ancient romans. They were encouraged to throw bones, cores, and leftovers on the floor like romans. Dormice rolled in honey and poppy seeds was one dish. Trojan pig - a pig stuffed with quail and beef parts was another delicacy. Pheasant and peacock brains, pike livers, and flamingo tongues was another. No one really cared. They were all so intoxicated they would've eaten anything. Wine, beer, and all the alchohol to drown a whale was served. He put a huge wooden bucket in his backyard labeled vomitorium so people could regurgitate there. After the dinner Creedo had his way with a few nymphs. Intoxicated with wine and some aphrodisiacs, the women were susceptible participants. He ravished them in a huge miasma of flesh, wine, food and lust. He felt better. This was a start. Home sweet home.

Chapter 10

The phone rang. A deep professional voice rung out. "The president would like to have an audience with Creedo next week. It has been scheduled. Would that be posssible?" I said of course and was anxious to tell the Senator. We packed our belongings and got on the Acela to Union Station. We stayed at a local motel near the beltway and anticipated the meeting between Creedo and the most powerful man in the world. We were both excited and a little nervous with anticipation about the outcome and what kind of subjects to discuss.

We spoke to a gray looking secret service man. He had a manifest with our names on it.

"You may proceed, the president is expecting you and your guest," the agent said, apparently not informed I had with me a man from the past.

"Thank you," I replied excitedly.

We entered the oval office slowly. It was smaller than I thought. Most corporate boardrooms were bigger. But in this room held more power and influence than all the palaces, legislatures, and foreign government capitals put together. The president was an average sized man, gray haired, and very personable. His presence filled the room. You could tell he had the world's problems on his shoulder, but he also gave the impression that he could handle it, as well as all the press criticisms and oppositon.

"I hear we have an authentic roman senator here," the president said with authority.

"Yes sir," I replied.

"I am Creedo, and it's a great honor to meet you Caesar," the senator greeted.

"Well you don't have to call me that. Mr. President will do. We live in a republic, there's no empire here," the president replied.

"With all due respect Mr. President, I beg to differ." Creedo shot back.

"So you're a senator from ancient rome; captured from a time machine. I heard from some of my assistants you have a lot of opinions about this place. Lets hear them. What is the USA like compared to the great roman empire," the president asked, not prepared for the extensive reply.

"Well, my english is not that great, but..." Creedo said interrupted by me.

"Nonsense, it's better than most Americans," I blurted out.

"First let me go over how Rome became an empire. Then I will explain how the

United States took the same path. Then I will explain the enemy you face in this, what you call the war on terror," Creedo explained.

"I'm anxious to hear your thoughts," the president stated.

"In 510 BC, Rome had been a monarchy for about 250 years under various kings: such as the Etruscans and other monarchs. The last king was Lucius Tarquinius Superbus, or Tarquin the proud.   Now he had a son, Sextus Tarquinius, and he was a bad prince. Sextus raped a noble woman named Lucretia. She was so upset about it she killed herself, but after notifying her relatives and friends. Her powerful friends and family led a revolt against the kingdom and overthrew Tarquin the proud, and he fled to Etruria. They then established the republic of Rome. Lucretia's husband Lucius Tarquinius Collatinus, and her brother Lucius Junius Brutus, became the first consuls of the Roman republic. The latter individual is rumored to be a descendant of the Brutus who famously conspired to kill Gaius Julius Caesar in 44 BC. So that's how it started. A small agrarian community that overthrew a monarchy, started off as a republic. The leaders would be two coequal consuls. Our checks and balances. Each powerful magistrate had at least one other who had to agree. Collegiality it was called. In addition to that assemblies met, every cives or citizens could vote; and a simple, but effective democracy was born," Creedo paused.

The president sat attentively. Staring at the senator he stated, "Go on."

"From this we grew like a flower. Rival tribes such as the Sabines and Samnites we defeated. Rome formed the Latin League. Like your NATO, it was used to acquire allies, and dominate the region. From the period of about 340 to 268 BC, Rome occupied and acquired most of Italy. In 283 BC, we defeated Pyrrhus at Tarentum.
We defeated the Greeks and Seleucid Empires. This battle for Mediterranean hegemony was successful. Then came our greatest rival Carthage. By this time though, Rome dominated the region, especially Italy, through allies, conquered city-states, and colonies. In 264 BC the first of the Punic wars began. It started over a minor conflict in Sicily at the time. Rome was allied with one side, Carthage the other. Carthage had a superior navy, the greatest in the Mediterranean. Rome had a superior army. At first the Carthaginians were winning because their generals avoided land battles and mostly fought on the seas. But the Romans developed an assault bridge on their ships called a corvus or raven that allowed Romes legions to fight naval engagements. Rome built more ships and won in 241 BC acquiring Sicily, Corsica, and Sardinia. The Second Punic war is famous for Hannibal and his elephants crossing the Alps," Creedo was cut off by the president.

"With all due respect what is the point of this extended history lesson," the president said tersely.

"You'll see please bear with me, just making parallels between empires. Anyway Carthage began getting its house back in order after that under the Barcid family and expanded its empire into Hispania. Rome was fighting the Illyrian wars at the time and couldn't check the power of Carthage. Hannibal attacked a Roman ally, Saguntum in Spain and the second Punic war began. At first Hannibal did very well. He led his troops into Italy and won the battle of Cannae in 216 BC by encircling a heavily superior Roman garrison and outflanking them by this method. With his elephants he crossed the Alps and was certain of victory. He defeated many legions, but assumed Roman client states would fight for him. They didn't and remained loyal to Rome. Meanwhile his brother

Hasdrubal was losing Hispania to the Romans, and Carthage failed to reinforce Hannibal in Italy. In Spain our great commander Scipio defeated Hannibal's brother Hasdrubal while he was attempting to reinforce Hannibal's army in Italy. Hasdrubal was killed in the battle of Metaurus defeating his forces. Scipio then invaded Africa and directly threatened Carthage. Hannibal marshalled his forces to Africa but was defeated at the great battle of Zama in 202 BC. Scipio was granted the title of Scipio Africanus and became one of the greatest Roman military commanders. After this peace came with a heavy price. You see Carthage was like the Soviet Union during your cold war. A rival power that threatened us for many years. So Carthage was stripped of its colonies, we imposed large war reparations, and of course another war was inevitable. After the second Punic War, Carthage started to rebuild. Still just a city-state now with no military, it experienced incursions from neighboring Numidia. They built up their military, paid off their war debt to Rome; but lost to Numidian forces. Disputes between Carthage and Numidia were resolved by the Roman Senate. The senate was partial to Numidia because they were a client state of Rome. Carthage now had a war debt to Numidia. Great leaders like Cato were stating 'Ceterum censeo Carthaginem esse delendam' or I also think Carthage must be destroyed, after each speech. In 149 BC Rome basically took a preemptive strike against Carthage, a helpless and powerless state that posed no threat to Rome and had no weapons of mass destruction. They demanded Carthage be destroyed and restored away from the coast of Africa, deeper inland. When Carthage refused Roman legions led by another Scipio, Scipio Aemilianus used shock and awe; sieged Carthage, burned it to the ground, and took those left as slaves after 3 years of fighting. They even sewed salt into the earth so nothing could be grown there. Basically Rome was a superpower that attacked a smaller state, that was no threat to them, for basically no reason than to assert their power in the region. History does repeat itself doesn't it? Now empires are always challenged, so Rome also was fighting the Macedonian and Seleucid empires during this time. The United States can certainly relate to this. Rome was fighting on many fronts. But finally Rome lucked out. King Attalus III of Pergamon bequeathed Asia Minor to Rome in his will. Rome was given huge amounts of territory by treaty. This was basically like the Louisiana Purchase of our time. However they still had to fight an upstart prince named Aristonicus for the territory, so war broke out from 133-129 BC. During this time powerful wealthy interests were amassing power and fortune through the Senate and lands conquered in other provinces. Large farms called latifundia were possessed by the patrician and equestrian classes squeezing out the small farms. Finally a pair of brothers named Gracchi enacted some reforms for the plebian classes. Attempting to put a limit to the size of farms and using his power as tribune, Tiberius Gracchus also tried to confiscate illegally seized farms and give them to the poor. He was assassinated by wealthy senators and became a martyr to the cause of the lower classed in Rome. His brother Gaius also attempted to limit the powers of the senators and pass social reforms. He was also assassinated. After this time about 400 years after the birth of the Roman republic, Rome faced many crises and terrors. The first came soon after the first century BC. General Marius who fought many wars against Rome's enemies, was rivals with a former protege Sulla for control of Rome. The basis for this rivalry was something about Sulla being sleighted about receiving credit for capturing some king in a former battle, but that doesn't really matter. Now anyway several wars were going on during this era. One was the Social War from

91 to 88 BC. This was basically a war over immigration policy. Rome's client-states wanted citizenship for their people. They had to pay taxes and supply troops for Rome, but didn't get any benefits. This would be like if Mexico declared war on the US over immigration policy. Rome eventually won the war, but gave in and made citizens of Italian states Roman citizens as well. In 88 BC an insolent King named Mithridates VI of Pontus butchered Roman citizens in Bithynia. Rome's senate gave Sulla command of the army to fight him. But Marius tried to usurp power by bribing the plebian assembly to give him command of the army. Not only was Rome fighting a foreign power but it was fighting between themselves. The Republicans and Democrats seen nothing like this. So obviously Sulla was pissed. He led his legions to conquer Rome. Marius fled to Africa. Sulla then went to confront Mithradates, meanwhile Marius with a fellow general named Cinna, led his legions back to Rome fighting bloody urban warfare for control of the city. Marius killed Sulla's supporters and took control of the city of Rome. Sulla won the war against Mithridates in 85 BC. Marius died of a brain hemorrhage soon after, but leaving his second in command Cinna in charge of the city. Cinna was a lousy general not loved by his troops like Marius. He was killed by his own troops actually, but the civil war was not over. General Carbo was in charge of the anti-Sulla forces, and Italy was ablaze with war and rampage. Sulla was in charge of the optimates, basically the patrician party. While his opponents were the populares or liberal, reformist party. The populares were allied with Rome's traditional enemy the Samnites as well. At the battle of Colline Gate, Sulla's forces won and slaughtered their opponents. This allowed Sulla's legions to control Rome paving the way for him to become dictator in 82 BC. Sulla was a rather harsh imperator or dictator of Rome. He imposed proscriptions or lists of people he deemed enemies of the state. If you were unlucky enough to be on this list, one was persecuted by the community, the property of the person and their family was confiscated, and you could be killed for a bounty. He also gave the senate more powers at the expense of the tribune assembly because the senate was controlled by the patrician class or optimates, the party of Sulla. One of the people he pursued in his proscription persecutions was a man named Gauis Julius Caesar. Now Caesar had family connections both pro and against Sulla. So he used his family connections to get him out of being killed by Sulla's forces during this time. In 79 BC. Sulla retired from public life and commented the in Caesar there are many a Marius. Shortly after this episode in our history came the Cataline conspiracies. Basically a social reformer with autoritarian impulses wanted to cancel all debts and assist the peasantry of Rome. He ran for consel and lost to Cicero his longtime opponent. When he lost his supporters tried to overthrow Rome and return us to a dictatorship with the help of slaves and plebes. In 62 BC he was killed by Roman troops enhancing the stature of senator Cicero. Shortly after this began the rise of Gaius Julius Caesar. Now as a Roman I tend to be long winded as you people say, I apologize for that I'll try to wrap this up. Anyway Caesar's career was great. He conquered Gaul and Egypt. Fought Pompey for control of Rome and became perpetual dictator in 44 BC. Of course he was assassinated on March 15, 44 BC in the Roman Senate."

"We Americans know all about Caesar; some of that history I didn't know. So I guess your point is that Romans got tired of fighting and gave in to more autocratic rule?" the president interrupted.

"Basically, do you know the history of how Octavian assumed power as well?"

asked Creedo.

"Yes, he battled Marc Antony and won at the battle of Actium in Greece. No need to elaborate on that," the president replied.

"Ok then. Well at least let me go over my interpretation on American history. I'll try to make it as brief as possible. Basically you started off like us as a small agrarian state. Like Rome, your country was ruled by an oppressive King, who imposed taxes without representation. Quartered troops in private homes, and disbanded duly elected assemblies; the King's actions inspired a rebellion that created your republic. Your pater patriae was George Washington. Father of his country he was called, which was originally a roman title bestowed on such eminent citizens as Cicero and Caesar. He established your great republic, and with other founding fathers came up with your constitution. After throwing off the shackles of monarchy, the United States like Rome pursued a policy of expansion and dominion. In 1804, your leader Thomas Jefferson bought the land called Louisiana Territory from Emporer Napoleon I believe his name was and doubled your territory. Shortly after that you fought the British again. They burned your capital down and you tried to acquire their land known as Canada. This failed but your quest for expansion known as Manifest Destiny continued unchallenged after that. In 1846 you fought Mexico. From this empire you acquired provinces known as California, Texas, New Mexico, and Arizona which expanded your nascent empire even further. I forget the exact reasons for this Mr. President but I think it was based on America sending your people to these lands so you could declare them yours. That policy reminds me of the saying 'Wherever a Roman conquers, he inhabits.' Anyway when your conquest of those provinces were complete, there was a great conrtoversy over slaves. Which territories should have slaves and which should not, dominated discourse until the great Civil War. After this was settled; slavery was abolished, the counrty reunified, your country could pursue its policy of expansion and empire. In 1893, some sugar farmers from America and missionaries went to the Kingdom of Hawaii to settle. A settled island with its own Queen, this band of intruders declared their own government. The United States supported this mutiny and acquired the province of Hawaii for your great empire. A few years later in 1898, the United States under the pretense of a naval ship that probably self destructed, invaded and acquired the territories Cuba, Phillipines, Guam, and Puerto Rico. After that your leader known as Teddy Roosevelt who liked to carry a big stick, swung his stick at Panama inspired a revolt so he could build a canal. This canal facilitated your trade between your east and west coasts. After a few years in your history came the war known as the great war. The germanic tribes united against your great power. In 1917 America with some allies such as England and France battled the great Germanic empire. You won and destroyed the great empires of Germany, Austria-Hungary, and the Ottoman empire. But the huns were not defeated yet. They regrouped and built up their phalanxes and navy under commander Hitler. He secretly built his army and with Caesar Mussolini in Rome tried to create an axis empire to rival America and England. This led to another war with germans. After America's second great victory over the vicious goths, hey we know how hard they are to fight, look at the battle of Teutenberg forest in 4 AD when we lost 4 legions to those rabble. Anyway the United States by glory of Jupiter, or whatever gods you have blessed upon you became the dominant power in the world. From the great weapon know as the atomic bomb dropped in 1945 your country ruled the world But soon after the germanic wars ended, a threat emerged from the east called the Soviet Union. Basically your Carthage, you fought proxy wars against this power in places called Korea, Vietnam, Cuba, and Grenada. You called this the cold war with nuclear weapons which could destroy the world you defeated this strange threat called communism. Finally in 1991, your nation won, and the empire of the United States led the world. During this time your politics had changed as well. The presidency which was originally supposed to be subserviant to the House of Representative and Senate (you people clearly desired to be the new Rome with that title) which was more democratically elected. But over time the power of this position grew with the power and expansion of your empire. It explicitly states in your own constitution that the Congress has the power to declare war. But like all caesars, he defines the meaning of words and laws, one previous leader even redefined the word is to his liking. Congress has not decared a war since 1941, but presidents have unilaterally as commander in chief fought and conqurered nations with a submissive legislature that acquiesces to the emporer by granting him powers not expressly granted in the Constitution. Your constitution gives Congress the power to declare war, it does not give Congess authority to delegate this power does it not? But like Rome under the great Augustus, the caesar defines his powers and leaves others the duty of justifying it.

So now your nation is fighting the Arabic peoples. I have some advice for you based on some books I have read."

"Let's hear your opinions. You seem like a very learned man. Hope you have some good advice," the president quipped.

"Alright, now let me ask you something, who was it who finally defeated the Roman empire?" Creedo asked smugly.

"Was it the Goths?" the president answered.

"No! Guess again," Creedo shouted.

"The Visigoths, Vandals, Huns. I don't know then," the president answered perplexed.

"It wasn't anyone of those people. It was Islamic arabs; the same people you are fighting today. Remember in the 4th century the Roman empire split in two, with the eastern half moving to the east with its capital in Constantinople. The western half fell in the 5th century, but the eastern Roman empire or Byzantium lasted until 1453 when Constantinople fell to the Ottoman turks under Mohammed the Second. So it was the Muslims, the same people you are fighting today who brought down the Roman empire." Creedo answered emphatically.

"Ok, that's very interesting. So do you think we can win our fight against Islamic extremism?" the president asked.

"First let me give you an idea who you're up against, what they're capable of, and historically speaking what to be prepared for. Now just a superficial look at history will tell you this struggle will not take a few years, or even decades. Remember the Muslims defeated the eastern Roman empire, the crusading armies of Italy, France, Italy, Germany, and England. That war lasted over 2 centuries and they won! They know this, they're well versed in history, apparently much better than you people and they see this as a new crusade. Now the only way America can win is by amassing a large enough empire to counter 1.3 billion muslims. You must unite the western world under an American imperium..." Creedo stated before being interrupted by me.

"Hold on. Look I love you Creedo but some of your ideas are wacky. I know how to avoid a hundred year war against the Muslims. We could end this thing tomorrow here's how. Make a grand bargain with the arab people. Say to the people of Waziristan or wherever Bin Laden and his henchmen are hiding, and say look; give us Bin Laden, Al-Zawahiri, Mullah Omar and all the Al-Qaeda leaders on our list, hand them over whenever you find these bastards, and in exchange we'll withdrawal all US troops from the Middle East, work on peace between Israel and the Arabs, send our troops home and work on energy independence. I mean with all due respect Mr. President I'm sure they don't like American troops on their lands. I mean how would we like it if say Saudi Arabian troops were in say South Dakota or something. I know they wouldn't like that! Make them a grand bargain, and let them know if we're attacked again we'll be back with vengeance. No one really wants a war that could last hundreds of years. At least consider it." I explained to the American leader.

"No, that's way too simplistic an approach. We need the forces there to preserve order, and protect American interests. Besides if we withdrawal now that would signal weakness to the enemy. We don't want the terrorists to think they can force us to leave." the president stated emphatically.

"I have a better idea. Unite North America under an American empire. Then acquire the South American continent. Have them pay tribute and contribute forces to your wars in the middle east. Europe and Africa could also join. They would also contribute force to the new American order. Australia could also be acquired being an english speaking country. With these new lands the United States would be a real threat to any Islamic caliphate or empire that would arise in the future. The United States must unite the rest of the world against any Islamic threat in the future. Trust me these guys are tough and should NOT be underestimated. Many great empires were buried in the sands of Islam I tell you," Creedo told the commander in chief.

"Believe it or not I like your idea Creedo. You speak from the wisdom and prescience of history. Guess we're going to need more troops!" the president roared out.

"Are you kiddin me! How can you actually listen to that proposal? Mine was actually sensible!" I screamed.

"In the long run we're going to need a much larger military in terms of manpower. They're going to have to come from somewhere," the leader said solemnly.

The chief of staff hurried in and abruptly but politely escorted us out of the oval office. I quickly glanced at the president smirking at Creedo as we left the threshold.

Chapter 11

Creedo and I took a brief tour of the US Capitol after our encounter with the president. We went with a tour guide inside the capitol building when Creedo looked up at the large fresco at the top of the rotunda. The guide explained it's titled *The Apotheosis of George Washington*. Completed in 1863 by Constantino Brumidi it represents the ancient Roman tradition of deifying a roman emperor by the process called apotheosis. This was done to Rome's founder Romulus when he became the god Quirinus. The fresco depicts many roman gods interacting with prominent Americans such as Benjamin Franklin, Samuel Morse, and Robert Fulton. It portrays George

Washington next to two female figures representing Victory and Liberty. They are surrounded by 13 females allegorically representing the 13 colonies. His arms stretched out, he's floating among the clouds into heaven. The scenes depict Roman gods and goddesses such as Minerva, the goddess of crafts and wisdom, pointing to a generator, a battery, and a printing press. Next to her are American inventors Franklin, Morse, and Fulton. Another scene features the Roman god Neptune with a trident, a crown of seaweed, riding a sea shell chariot drawn by sea horses. It also depicts Venus, goddess of love, pulling a transatlantic cable. A scene representing commerce portrayed Mercury, Roman god of commerce with his winged petasos, sandals, and a caduceus, which is commonly known as the symbol of medicine; is shown giving a bag of gold to Revolutionary war financier Robert Morris. The Roman god Vulcan, god of fire, is shown with a cannon, and a steam engine in the backround. Finally, Ceres, the goddess of agriculture, is shown with a wreath of wheat and the cornucopia, symbol of plenty. She is sitting on a McCormick mechanical reaper, a farm tool commonly used in the 19th century. Young America is represented by a man holding the reins of a couple of horses, while the Roman goddess Flora gathers flowers. The scenes are used to represent science, the seas, commerce, agriculture, and mechanics. What is strangest about the concept however, is probably the representation of an American president becoming a god. It seems counter to the perception that presidents are first among equals. The concept of the divine right of kings was something our founding fathers fought against, so the portrayal of George Washington becoming a god seems odd as the fresco at the center of the US Capitol, symbol of American democracy. The painter Constantino Brumidi, worked in Rome at the Vatican, and I suppose got his inspiration from the church and Roman theology. Creedo absolutely loved it. He thought it showed the link between our empires better than anything else he'd seen before. Right at the heart and the center of the US Capitol was a painting with the founder of our nation being deified in a Roman process, surrounded by Roman gods with American citizens. He truly realized at that moment he was in the new Rome.

Jim Douglas had been with the NSA for several years. A bald, paunchy man in his late 40's he loved his work and his work loved him. Mostly assigned to terrorism cases he was surprised to hear his new assignment involved confiscating a time machine unit near the deserts and canyons of Sedona, Arizona. He came with a team to our cavernous facility teaming with scientists and engineers. Some of the best and brightest worked there. The NSA team raced their black cars there around 2 AM in the morning. They brought huge tractor trailers to haul off the unit. They raced into our facilities screaming something about this being the interests of national security, hauled off the displacement unit after dismantling it into pieces and carried it off to their secret facility in Virginia. Jim Douglas was never involved in a project like this. Stealing a prize possession from peaceful scientists for government purposes was nothing new to him. But an actual time travel device excited and frightened him. He thought about the possibilities riding the long trip back to Virginia. His partner was Stacy Fields. An operative with the NSA for about 5 years, she was in her late 20's, blonde hair, blue eyes, and athletic build. Looking something like Jodie Foster in Silence of the Lambs, she was the archetype of what a female agent should look and talk like. She had no doubts of what she was doing. Doing her job and serving national interests was enough for her no matter the duty, no matter who had to be terminated.   What they were not told of is the reason for this assignment. They never were. Just doing their duty and following orders was all they knew. When they arrived in Virgina, Douglas got a call from the President. He never talked to the president before, but he was not surprised he knew this was important. His cellphone rang.

"Hello Jim. This is the president. Good Work! Thank you for your years of service to the United States. This machine will provide us with troops and personnel for years to come in our struggle against terrorism and a more peaceful and stable world," the president stated.

"It's an honor to talk to you sir. Thank you. I do my best for you and the United States of America!" Jim proudly said.

"It's great to have someone like you work for us. Keep up the good work," the president replied.

"Of course. I hope you have good use for this unit," Jim asked.

"Oh we do, believe me we do," the president stated emphatically.

Chapter 12

"However our present interests may restrain us within our limits, it is impossible not to look forward to distant times, when our rapid multiplication will expand it beyond those limits, & cover the whole northern if not the southern continent, with people speaking the same language, governed in similar forms, and by similar laws." Thomas Jefferson speaking to James Monroe 1801.

"They took it. The bastards took it," I screamed.

After getting a call from Sedona lab, they explained how federal agents seized our time machine for god knows what. I was livid. So booking the nearest flight to Sedona Airport, I rented a black Chevy Malibu sedan and headed off Rt. 89 to discuss the situation with my scientists.

"So what the hell happened," I asked impatiently.

"Several men and a woman came in, presented their government badges and took the time displacement unit. Hauled it off in trucks, and that was it!" One of the particle engineers volunteered.

"Ok, we'll get it back I promise. I have friends in high places. I just talked to the president of the United States. Creedo and I had a very thoughtful and enlightening discussion with him on history and politics and stuff. I left him in a hotel in DC. I'll talk to his people and see if I can get it back. This things too dangerous in the wrong hands and I especially mean the government!" I emphasized.

"Good luck. I don't know what's going on," the scientist queried.

"I'll find out. And I'll get back to you as soon as it's resolved," I said tersely.

I got back to DC. We stayed at the luxurious Hay-Adams hotel. A stately lobby, temporary residence of dignitaries, senators, and visitors to important people. I made a cell phone call to the chief of staff of the president.

"John Balducci speaking. How may I serve you?" he answered in a professional, baritone voice.

"What the hell have you done with my time machine asshole!" I shouted.

"Calm down. This is a national security matter. I can't discuss this," he replied curtly.

"Why not! I need to speak with the president." I demanded.

"He cannot discuss this matter. It is a classified matter. Trust me, your time displacement unit is in good hands, and will be used for good purposes. Even that was saying too much, and you didn't hear it from me. That's all I will say on the matter." He hung up quickly.

I frantically called all my contacts with the government. I went to Langley and demanded answers. All I got was a constant barrage of can't discuss this, national security, and classified information. The government had my time machine and what they would do with it I thought god only knows.

The NSA facility in Virginia was teaming with anxious scientists, engineers, military head honchos, and of all people Latin professors and Roman scholars. They were prepared to assist or embark on the expedition of a lifetime. The plan was to go back in time; to the height of the Roman empire, seize Roman soldiers, teach them english, and train them with modern weapons to fight America's wars. They needed Roman scholars to approximate where these legions would be at a given time. They also needed Latin scholars who know the language as interpreters, so they could communicate with them and teach them english. They also needed a small arsenal so they could initially impress them by force. If this worked this meant an unlimited supply of troops for the world's greatest military. Enough to crush any insurgency and fight any wars against any opponents. The military had been stretched too thin. Relying on the reserves and the national guard, there just weren't that many new recruits in this all volunteer army. The vast legions of ancient Rome would make a perfect recruiting tool for the new imperium. Jim Douglas and Stacy Fields were ecstatic to lead the team.

Douglas and Fields were ready for their mission. Aboard one of the largest military cargo planes in the world, they contemplated their roles in capturing Roman legionaires for the US military. Douglas, a very professional agent, discussed the ethical issues involved in basically stealing a time machine from MIT scientists for military purposes. Not having second thoughts, he really wanted some reassurance from Agent Fields that this was the right thing to do.

Flying at an altitude of 35,000 feet over the north Atlantic, Agent Douglas looked wistfully at Agent Fields. He casually jerked his head.

"Everything ok with you about this mission?" he asked with the slightest hint of a smile.

"Absolutely. I'm gung-ho. And you better be too!" she exclaimed.

"Yes, me too. But I guess I'm trying to say. Did you have any qualms about stealing a time machine from people. I mean just grabbing it from scientists like that?" Douglas asked hesitantly.

"No! And you better not have any reservations either. Look we've been training for this a long time. Now's not the time to get cold feet. We did this for national security. We are going to impress Roman soldiers so American soldiers don't have to die. What more noble and worthy mission could you ask for. We are doing this for America and our troops. If we had to take it, so be it!" she answered passionately.

"Ok. I just hope everything works out like we plan. I hope the Roman soldiers don't fight us too aggressively. I mean imagine seeing people from the future trying to capture you. I mean what would you do?" he inquired to agent Fields.

"We have enough nerve agents and chemical gas to knock out these people for days. If they try to fight back we got enough tranquilizers to incapacitate hundreds of legions. You got nothing to worry about. When they wake up, we'll remove their primitive weapons of course. We also have a special substance that when injected makes them susceptible to our will. Makes them willing to fight for us and open to suggestion. The NSA has everything covered," she explained with a steely gaze towards agent Douglas.

"Ok, great. Hey I never doubted our mission or anything. Just needed to know if you were ok with this," he stated glancing at her.

"Of course Jim. I can't wait for this!" she stated in a way that quashed all doubts.

Chapter 13

I was pissed. They took my time displacement unit. I wasn't really surprised, but fortunately I had a spare unit to use in case of an emergency like this. I could only speculate what they were going to use it for. Military uses I surmised. I wondered how they found my time machine and who told them. There was a spy in our midst. Someone on our team working for the CIA or NSA who tipped them off. Bastards! Now I was the only one who knew where the other unit was and how to use it. It was 500 feet below ground, in Sedona, near where the time displacement unit they took was located. What the NSA fortunately couldn't detect was a microscopic GPS unit on the time machine they took. The race was on. They were well armed, but so was I. I didn't want to kill or hurt anyone, but I felt a moral obligation to stop them. I would see where they would take my machine, and with the resources from my MIT grant, could intercept them. Before flying from BWI airport to Sedona I had another conversation with Creedo who was in Boston. He was grading papers in his office, when I called him on my cell seeing if, of all people he could give me some advice.

"Creedo, this is Mike. What the hell do you think the government wants with my time machine?" I asked impatiently.

"Your caesar was very impressed with what I had to say to him wasn't he?" Creedo stated confidently.

"Yes, I suppose he was. But what could he want with my machine?" I expressed sharply.

"The president said something about needing more troops after my speech to him, what could that mean," Creedo propositioned.

"Yes, with the wars in the middle east, the military needs troops since we don't have a draft anymore. You may be onto something," I said excitedly.

"Maybe the stealing of your time device is related to the need for more troops?" Creedo submitted.

"That could very well be it. But what could they be up to?" I wondered.

"I don't know. But I hope he takes my advice to expand your empire. Or at least be more honest about it. America's days of being a simple republic is over. I just read in one of your defense publications that your country has over 700 military bases in 130 countries. Maybe even as much as 1000 bases, because many are secret. With all these bases more troops are needed to police the world," Creedo informed.

"Yes, but I want to figure out how time travel relates to the need for troops, if this is their reason; and what they're going to do to get more troops. They could also use the machine to acquire gold or treasure," I said quizzically.

"Well, that's my theory anyway. Look don't worry about it so much. I'm sure your government knows what it's doing. They're probably doing the right thing for national security and the protection of the republic," Creedo reassured.

"Hell. Not much I can do about it. I'm sure they're well armed, not much chance of getting the time displacement unit back. I just hate the fact that the NSA and CIA can do anything they want. They even kidnap people in Europe and send them to countries with no human rights protections like Egypt and Syria. And now they took my machine! Just pisses me off that's all," I said solemnly into the phone.

"Welcome to the empire of the United States," Creedo chuckled.

I turned off my cellphone and started speculating. I could track them down but there's no way I could overpower them militarily. This was the most powerful government in the world I was dealing with. I couldn't even rely on my GPS unit because the GPS system is run by the US Air Force. The GPS system was originally a set of satellites launched by the USAF to provide missile guidance for the US nuclear arsenal. Its 28 satellites determine one's location by a complex system of determining the time zone location of the satellite in coordination with the location of the other satellites in conjuntion with the time of day and cross-signals with the other satellites. So all I could do for now was secretly monitor the location of my time displacement unit by moving around and hope the Air Force wouldn't distort the signal or detect what I was doing.

Chapter 14

Creedo sat solemnly in his office in his black plush swivel chair, casually shaking back and forth. He was a popular teacher. None of the students knew he was from ancient Rome of course. As a cover they were told he was from Italy; Rome of course, but they had no idea who he really was or what he was doing there. They didn't really ask. All they wanted was to pass, so they didn't ask too many questions. It was perfect for him. A real ancient Roman senator teaching Latin in its original tongue. He was thinking about his family. He thought about them from time to time, but didn't want to go back. I offered him many chances to go back to his own time to be with his family and people, but he enthusiastically wanted to stay put. He really loved living in America and felt a sense of purpose and calling even more so than in ancient Rome. He told me his wife's name was Agrippa, she was 31 years old. Flaxen hair, sea blue eyes, and razor sharp wit were some of her traits. His sons were 12 and 16, they were healthy lads who loved to swordfight. Their names were Brutus and Tiberius, and he wondered what became of them after coming to his new time and place. But he still wanted to stay here, he insisted on it. A small, young man darted in. A frosted blonde MIT grad student, I sent him to retrieve Creedo and meet me in Sedona.

"Creedo, Mike wants you in Sedona!" the young man exclaimed.

"I'm busy now. Tell him I'm staying here!" Creedo yelled back.

"This is important. I think he needs you!" the student insisted.

"Whatever he can do, he can do without me. I like it here, and I don't think he should be fighting his own government!" Creedo replied.

"He'll be disappointed, but if that's your decision," he sighed.

"It is. I'm staying here for a while. I like my life at this school, and the students love me too," he stated, partially referring to the multitude of orgies he had with them.

"Okay, I'll give him the message then!" he shouted in a huff.

The crimson sun gently kissed the english bluffs as it rose into the billowing clouds. Fields and Douglas were in the field waiting for further instructions. They were about 300 meters from what had been the northern most fortification of the Roman Empire. The great palisades known as Hadrian's Wall was built there 1900 years ago.

"To separate the Romans from the Barbarians," was the purpose for its construction according to Emperor Hadrian. In 55 BC Julius Caesar was the first Roman to make an amphibious landing in what is now Great Britain. In 43 BC Emperor Claudius sent Aulus Plautius with 24,000 soldiers to subdue the Celtic tribes and establish Brittania under Roman rule and occupation. This went rather smoothly but the Scottish clans did pose a menace from time to time. By the time of Emporer Hadrian in 122 AD the Roman empire was no longer expanding, reaching its zenith under Emperor Traijan a few years earlier. To fortify his boundries of the northernmost part of his empire he had his troops- skilled in carpentry, masonry, and craftsmanship- build what is now the most important Roman era structure in Britain. At 73 miles long it took about 8 years to build. When it was designed it was about 10 feet wide, and up to 20 feet tall in some sections. Started from the east at the river Tyne and finished in the western part of England at the Solway Firth it stretches from one coast of England to the other. About 10 miles south of Scotland the wall is entirely in the England of today and used to guard against Scottish and Celtic clans and tribes. The giant palisade also was guarded by about 20 forts, each with what is estimated today to have occupied 500 to 1,000 troops. Constructed mainly of stone, certain sections were made with turf that proved not to have the structural integrity. Every so often the wall contained milecastles, stone towers that held two turrets-with between 8 and 32 soldiers each. Also, the wall had additional ramparts with mounds and ditches to the north and south of the wall, as well as roads bridging the declines granting the soldiers access, while deterring invaders from the north. It was used and occupied by Roman garrisons until 410 AD when Rome withdrew from Britain, but it was still used by English forces after that. Built by the legions themselves, wall detail was almost recreational duty for skilled Roman soldiers whol built and maintained the wall, it's roads and fortifications for many centuries.

The NSA and CIA collaboration sent teams of field units to the town of Carlisle in the northwestern part of England. Part of a roman settlement known as Luguvalium; latin for wall of the Lugus, remnants of Hadrian's wall run through the Stanwix sections of the town. The teams were split up into operational field units headed by the CIA, reconnaissance units headed by the NSA with Fields and Douglas at the helm, as well as a communications unit with NSA agents-being their specialty, and a command and control unit that would give orders and cover their tracks. All this was in place by the time Fields and Douglas got there. Preparation took weeks by the best and brightest in the intelligence community. The time displacement unit was filled with all necessary equipment to do the job. Tranquilizer darts and pistols, incapacitating agents such as CS and CR gas canisters ready for aerosolized usage, and secretly acquired from Russia; the potent Kokokol-1 gas which could knock someone unconscious in seconds and last 2 to 6 hours, which was more than enough time to subdue someone. They were prepared, they launched operations to take Roman soldiers from the wall for use in US armed forces. Their plan was to take these legionnaires from the forts and the milecastles of the wall a few at a time but not enough to jeopardize Roman stability and control of the northern border of Brittania from the Scottish tribes. So a few hundred soldiers were to be taken every year from about 122 AD to 410 AD. Troops could also be taken from other fortifications in Roman history, as well as other periods in time. An unlimited source of military might and power for the new imperium.

The interior chamber of the cavernous streamlined time displacement unit could hold 150 people and safely transport them through time. The US government secretly contracted to build a nondescript airplane hangar in the Stanwix section of Carlisle. Technicians were preparing to set the ETA for 135 AD in a test run to impress roman legionnaires for the military. Inside the hangar smelled of steel and gas. Fields and Douglas were ready. They were psyched, so to speak, and ready for duty. With an acute look Fields ever so slightly winced at Douglas as they embarked on the greatest journey of their lives.

Chapter 15

The machine trembled with tremendous rage. It felt like being projected a thousand miles an hour while standing still. The gravitational forces were difficult for the team to bear. When they felt like it was more than they could possibly bear it came to a sudden and long anticipated halt. Fields shook her head as if reawakening from a surreal dream. She gently poked Jim to resurrect him. Jim shouted. He was conscious. She thought *thank god.* They were to go out and survey the wall's position and the easiest spots to capture soldiers without being attacked.

They exited the massive steel and titanium door with some degree of trepidation and excitement. They were on a mission of course, but they were time travelers. The suspense was palpable. They looked around with their infrared goggles. A team of ten agents would scout the area. It was dead of night. The time was 10:48PM March 12, 135 AD.

"Follow me I know where I'm going!" Fields said bruskly.
"Yes, we have no satellite GPS now, cuz that's about 2000 years in the future," Jim replied nervously.
"Quiet!" one of the team members whispered bruskly.

Fields raced passed the other members of the team. A little too anxious, she raced towards the mound that constituted the first defense of the wall. They all had the most advanced infrared goggles available. They started to crawl now in black camo fatigues, invisible and ninja-like in their movements. With great stealth Jim spotted a soldier about 50 yards away. Instinctively, he got out his sniper rifle and pinpointed his target. For a split second the awe of being the first modern person to spot an ancient Roman soldier entered his mind. Then he immediately locked on his target, seeing the Roman soldier was heavily covered in body armor there was not a lot of surface flesh available. The Roman guard would not make it easy. He could only make a headshot, but that might kill him. So Jim put the sniper rifle down and decided to use the gas canisters instead. But they'd have to get closer, much closer.

Douglas signaled to Fields to come his way. They would flank to the east of the milecastle. Their aim would have to be perfect. They crossed the first mound, and then the first ditch. One of the men was holding a gas-canister launcher with Kokokol-1 cartridges. Jim grabbed it from him and aimed directly for the legionnaire. The canister hissed angrily over the wall, missing the target by a few inches. The noise startled the guard. He motioned for the other guards to come to his loocation. He looked north, west, south, and finally east. The NSA team immediately split up. But despite all their valiant attempts at stealth; Jim looked up at the top of the wall - *spotted*! A spear violently came down from the wall like Jupiter's bolt and missed Jim Douglas by inches. Fortunately for him a ditch was nearby which he propelled himself into. They couldn't see him now. The soldiers on the wall were preparing a catapult to rain death upon the crew. Jim spotted the milecastle, and just before the catapult was about to spring, fired the canister at its base. He squeezed with all his life and aimed the gas toward the men. Direct hit! The soldiers went down, they were momentarily safe from peril. Douglas and Fields told one of the men to get an extension ladder on the two-way radio they were carrying. He grabbed it and used it to scale the wall. They came to the top and saw there was about 30 men to grab. All unconscious from the gas, they would remain so for a few hours. A few men from the team put them securely in shackles and tied them. Then they took them in the time machine. The process of military renditions continued. Fields and Douglas needed to move to another section of the wall. They needed about 100 more men for this to be a successful mission.

The next section would prove to be easier. Fields and Douglas and the eight others would get a break. The soldiers were apparently having a late meal and being in a section of the wall where no incursions were likely, let their guard down a bit. Having found them a bit cumbersome, some of the soldiers removed their leg armor before mealtime. Douglas used this to his advantage, targeting their legs with his sniper-rifle and knocking them out. They never saw what was coming, and even if they did, they would never understand it. The team was getting better at their craft. Like a high-tech cat and mouse game where their prey had no chance. They picked off their targets one by one. Douglas preferred tranquilizer darts to gas, in that the gas was unreliable, subject to the wind, possibly fatal, and less direct. Besides Douglas was a marksman, fluent in all weaponry and targets - moving or stationary. This was his game and he was playing it well.

Douglas and Fields engaged in a quick smile with each other. They were almost done renditioning the roman troops for future assignment. They were down to the last few men they needed to fill the unit. They tranquilized the remaining men and the unit was full. All unconscious and tied with straight-jackets made with the strongest material; they were ready for their trip to present day England. Hesitant to experience the travel to the present Fields and Douglas and the rest of the team did their duty and sat motionless in their assigned places. Fields slowly smiled to herself and thought "mission accomplished."

Chapter 16

The door was open to Dr. Brown's office. I entered the room anxious to explain what was done and ponder possible solutions. He was pleased to converse with me, especially since my falling out with Creedo over his refusal to help me get my time machine back from the government. The professor stretched out his back to receive me while I stood.

"The bastards took it Harry, what do I do now?" I demanded.

"They think they can do anything. Hell they lie about going to war, they certainly think nothing of this," he responded.

"Ok, but what can I do about it. Can you think of anything?" I wanted to know.

"Not much I can think of...unless you have a spare army in your basement you're not telling me about," he surmised.

"No not that. But I can track them," I revealed.

"Of course, you got a GPS tracking signal. Don't you think the government would've found that by now?" Dr. Brown quipped.

"Yah, but this is invisible to usual detectors, this is a state of the art GPS unit that I believe...well hope can't be detected," I stated.

"Good. And I'm not surprised Creedo refuses to help in any way. I never liked that guy. I respect the fact that he's an ancient roman, but that's about it," he volunteered.

"That's okay. I got some of my friends at MIT who are on this. Hey. this might sound like a silly question, but would you like to go to England with me?" I propositioned.

"On a futile mission to fight government forces to save the future and all mankind? Count me IN!" he exclaimed.

"Great! We need to go soon. I'll email you the details," I forwarded.

"Hell no Mike. The government monitors that. We'll meet somewhere, and only exchange messages by human relay. You're smarter than that!" he scolded.

"Of course. I was just kidding Harry. We'll set up the operation soon," I shot back.

"By the way where in England are we going?" he questioned.

Sensing the room might be bugged I whispered ever so slightly in his ear,"*Carlisle*."

Back at their temporary field base, Douglas and Fields were about to be debriefed on the operation. The op was a success and would be used as a prototype for future operations of this kind. From now on the government would use the military to rendition the roman soldiers from the past. They also built a covert installation to house, train, teach english, and acclamate the troops to modern ways and weaponry. Once this was done they would be assigned an American sounding name and sent to a US military base around the world and sent to combat. This would ensure American troop strength for the forseeable future.

"We did it!" exclaimed Fields.

"Yes we sure did," replied Douglas. "Now the Pentagon can take this over. Thank God."

"We can still help in communications from the ground and in present time," quipped Fields. "I still want to be a part of this."

"Me too," affirmed Douglas. "But I can't wait to go home and rest for a while, that experience was taxing."

"You know it!" said Fields.

They took the congratulatory call from the president. The accolades poured in from the higher ups. When Douglas was alone with no one in sight, he pulled out his cell phone. He dialed cautiously.

"Creedo, thanks for your help..."

Creedo was done with all his classes for the day and was ready to unwind in his office. He slowly leaned back in his chair contemplating the reasons for his assistance in the roman acquisition program and the impoundment of the time displacement unit. He told them the location of the time machine, and offered advice on roman troops and logistics in the ancient world. Without him the operation would not have been successful. He thought he was to being loyal and patriotic to his new empire. He was glad to assist any way he could. He then realized his civilization gave birth to America. It was discovered by an Italian, Christopher Columbus. It's named after an Italian too, Ameriga Vespucci, the first explorer to name the new continent. He would help America transition in the new phase of the American empire.

Chapter 17

Dr. Brown and I met up at South Station in Boston. We discussed our future plans and arrangements. We would drive cautiously to Sedona in his car to apprehend the spare time machine before the government would. We needed to package, ship, and transport the unit to England without being detected by authorities. It was a delicate operation, but it had to be done. I brainstormed with my scientists how we would go about doing this. We disassembled the unit and labeled each piece scientific or medical equipment and got the proper paperwork to verify it. We then numbered each parcel and sent it to several trustworthy colleagues in England who were instructed to hold on to the packages until further notice. Dr. Brown and I then went throught the process of legally changing our names and identities and took a flight to Carlisle Airport in England to see what we could do.

Once we arrived, we rented a flat in Carlisle under our assumed names. We also needed to find a large storage facility to reassemble the time displacement unit, that would be at once spacious and discreet. I rented the flat from a kindly, middle aged

English woman who presented me with the key on arrival. I entered the room and viewed the normal amenities a small apartment should have. It was a duplex inside a small townhouse it contained a 4 foot high cabinet, a queen size bed, bath, kitchen, and all the normal amenites of an english flat. I spotted a green and black box with a padlock under the bed. It had a yellow post-it on the top stating "YOUR KEY WILL OPEN ME" so I used the apartment key to open the padlock. On a standard sheet of office paper it had the words WHEN THE RAILS' CLOCK STRIKES THREE UNDER THE EAST ARCH BE written in black crayon. Dr. Brown and I didn't know what to make of it, but we decided it must be important. The closest train station to Carlisle was the historic Newcastle Station in Newcastle upon Tyne. On the other side was written in green crayon BURN THIS PAPER IMMEDIATELY. Neither one of us smoked, but there were matches in the kitchen for lighting the gas stove tops, so we used one of them to incinerate the paper.

The next day we arrived at the Newcastle Station under the east arch. We strided the parking area and saw a homeless man with outstretched arms. He wore a crumpled jet-black robe with a canopied hood. He was a middle aged man, dirty blonde hair, with a 5 o'clock shadow. The impression I got from him was not of a destitute man, but a professional; possibly in law enforcement. "Give me a dollar, and I'll give you information," he brusked. I did what I was told. He gave me a yellow envelope which I quickly pushed into my right coat pocket. "What is your name," I asked. He said, "Let's just say my name is Lou Parn. Names aren't important just keep walking and quickly!" he ordered. I complied and kept walking with Professor Brown. The professor turned to the man and stated softly under his breath, "Leave the riddles to Dan Brown." Then we swiftly took the transit to get home and read what was in the envelope.

When we arrived at the flat we quickly shut the blinds and drapes as well as locked the door. We turned on a light situated on a brown desk in the living room, opened the yellow envelope and read the letter inside. On a plain yellow sheet was written: I KNOW WHERE YOUR TIME MACHINE IS. DO NOT ATTEMPT TO RECOVER IT NOW. THIS WOULD BE A SUICIDE MISSION. I WILL GIVE YOU MORE INFORMATION LATER. We had no idea who wrote this letter. His name certainly wasn't Lou Parn. We didn't know who he was, if he could be trusted, whose side he was on, or if he was working for an enemy government. But our curiosity got the best of us, and we didn't have much to lose. So while we didn't completely trust this stranger, we would do what he said. Information was worth any price.

The parts came piecemeal by FedEx to our contacts in England. We rented a large barn in Cumbria county, on a nearby farm, to reassemble the time displacement unit shipped from Arizona. Dr. Brown and I figured we would do some time traveling in order to gather clues what the government was doing. We figured that since the military took the time machine to Carlisle; site of Hadrian's Wall, it must have something to do with that. Perhaps they wanted to build a new wall. Maybe they wanted to change the past for some reason. We could only speculate at this point.

"What do you think they're doing with our time machine Harry?" I inquired.

"I don't know Mike, Do you think we can trust this Lou Parn guy?" Dr. Brown cooly responded.

"It might be a trap; lets be on our guard, but lets see what information he gives," I said.

"Yeah, what else can we do right? I mean what choice do we have?" he shrugged. "Let's play it by ear," he concluded.

"Why do you think they came to Carlisle?" I brought up.

"Must have something to do with Hadrian's wall; don't know what exactly, but that's what I'm thinking. Maybe your friend Creedo's got some ideas. I know he's been unhelpful lately; but it's worth a shot. He's from ancient Rome, he's got to have some suggestions. You want to call him back at the college?" he suggested.

"No, that guy's been acting strange lately. Don't know what's up with him. We need to do this ourselves. Sorry," I concluded.

"Fine...let's start by reassembling our displacement unit. Check out when Hadrian's Wall was first built."

We both nodded and went to work.

Chapter 18

The reassembly process took a few weeks. We recovered the boxes from several FedEx centers in Cumbria county. The dairy farm was spacious enough for us to work in peace. We did research on the history of Hadrian's Wall. When it was built, why it was built, and who built it; were researched thoroughly. We would have to be prepared for the travel and so brought flak jackets we got from an Army/Navy store back home. We had night vision goggles, and camoflage wear to hide ourselves and didn't want to engage in any confrontations with the indigenous population. This was a fact finding mission. We needed to find out what was going on.

Dr. Brown and I prepared the time unit for activation. We chose the year 140 AD to transport to (near the Stanwix part of the wall) to observe anything unusual going on. The two of us wouldn't venture outside but would use our audio and video sensors as our eyes and ears in the field. We would position ourselves about 100 yards from one of the famous milecastles, used to garrison the wall with paramilitary units from the Roman empire. What we found (or didn't find) was shocking to say the least.

It was a damp, dreary May afternoon. The gray breeze made us happy to leave. We set the ETA to May 16 140 AD--almost 1900 years in the past. The launch went smoothly. This was Dr. Brown's first time travel experience, so I assisted him with the process. When we arrived we turned our sensors on and surveyed the situation.

First the TV monitors showed it was early afternoon around 2 PM. It was a much brighter, warmer day than we had left. The time displacement unit could withstand any attack from the troops-it was both bomb resistant and bulletproof. They had primitive spears and arrows to deal with no match for our titanium armor. We looked around and made our observations. Dr. Brown manipulated a hand-held joystick to control the camera. When we arrived we were at the base of the wall looking vertically at the apex of the milecastle.

"Focus on the top of the wall. Let's try to find the guards," I stated.

"You know there should be a few dozen troops on the wall and in the houses but I don't see any?" Dr. Brown observed.

"You know you're right! Why don't I see any troops that's weird," I said, while he panned horizontally.

"Panning maximum range all along the wall, lets see if we can find anyone," he replied.

"Damn, still so see anyone. Wonder what happened to them?" I shouted.

"You know if we don't find any troops, this means we should look around and see if we can find anything," he suggested.

"Yah, but let's keep looking, I want to make damn sure no one's there if we're to venture outside," I cautioned.

"I'll keep panning, but I really don't see anything," Dr. Brown concluded.

The monitors still showed no movement or life. We were perplexed. Both of us expected scores of roman auxillary garrisons on the wall, but there was nothing along this section. In a few moments we would open the outside hatch on the starboard side of our unit. We were scared, but excited. When we opened the hatch, the air knocked our lungs with a clean deep breath. This was air not polluted with hundred of years of coal, gas, and carbon monoxide like our own time. This was air cleaner than any modern man will ever know. After getting accustomed to this, we made our trek, and gathered our facts for the task at hand.

We walked a few yards from our displacement unit towards the base of the wall. A few feet from a milecastle we observed nothing out of the ordinary. But we still observed no signs of life. I then spotted a flicker of movement about 15 yards from where we were standing. It appeared to be a paper or leaf gently massaged by the wind. We ran closer to it, and identified it as a papyrus of some kind inscripted with latin. We walked along the wall and still found no troops guarding the section we observed. Not wanted to press our luck any further, we walked back to the time machine with the paper to discuss our next set of actions.

"Hey, what do you make of this? I mean no people. And this parchment. We'll have to give this to a latin professor for translation. It might give us a clue why there's no troops here," I explained.

"Yah, this thing seems too strange. I just feel like history's been tampered with. Like the government had something to do with this, do you get that feeling too," he stated.

"I sure do. Let's get back to our own time, and call Boston to get this translated. This may reveal something," I suggested.

"OK," he concluded.

Chapter 19

They needed to cover their tracks. The NSA-CIA team couldn't significantly alter history--or be caught. Therefore when they took soldiers from Hadrian's Wall they couldn't have the northern boundary of the Roman empire overrun with english or scottish tribes. So they had a plan. When they took troops from the wall they brought

Creedo along. He told the commanding officers in charge of fortifications never to speak or write of this under penalty of the gods. After the legionnaires were taken from the wall, they would be replaced by audio-animatronics covered with realistic appearing latex made to look like roman legionnaires. They would move and were so realistic they would fool the tribes into thinking the wall was still well guarded. They could even speak--for extra realism--in latin of course, to really fool any would be invaders. These were new state of the art animatronics that needed no human operators. They even had basic AI that contained motion sensors that could defend the wall, and were equipped with realistic spears, shields, and tunic that could deceive even modern man. The real troops left were not to speak or write of this. Most of the actual troops were left to guard the wall of course. But the robotic troops were put on sections of the wall to fill the gaps. When the roman empire fell, the government team took the animatronics away; removing evidence that a more advanced civilization was there.

The ancient romans were properly deceived. Believing that the US military were sent from the gods, sent what they thought were movable statues, continuing to reinforce the wall. The actual troops taken however needed to know the truth eventually. Creedo assisted in this process, notifying them of their new country, place, and time. Most of the troops were of course perplexed and angry. But after being well fed, learning the new language, and taught about the latest weaponry, they adjusted nicely to their new situation. In a coffee shop in northern England, Fields and Douglas were musing.

"This operation was really successful. I've been thinking; why use roman soldiers, you have to teach them english and that takes time. Why not use english soldiers from the past, so you don't have to teach them a new language," observed Douglas.

"This is the US government you're talking about and that would make too much sense," Fields replied.

"Yah, you got that right!" he commented.

"But seriously, between you and me, I hear phase II will be exactly that," she whispered.

"The government doing something sensible, heaven forbid," he chuckled.

"Yah, the british empire was pretty great in its day, the sun never set on the union jack right," she stated.

"There's really an unlimited amount of manpower and resources we can draw from the past, the most powerful thing in the world is time. It's now the greatest weapon," he shot back.

"Yah, it's great huh," she yelled.

"Yah, but it's too bad an invention as wonderful as this couldn't have been used for more peaceful purposes," he sighed.

"I just hope they call me again, I'll be ready to go!" she finalized.

"Me too Stacy, me too," he concluded. Before they left, Stacy motioned towards her left holster. It was the latest Browning pistol; top of the line. "Here before you go I want to give you this," she said, "It'll never let you down." He thanked her and departed.

The ancient papyrus was being translated in Boston by a Latin professor Dr. Brown and I both knew and trusted. It seemed to be a journal entry by one of the sentries on guard. It translated from the original latin as,"We are not supposed to write or speak of this, so I shall toss this over the wall as far as the winds shall take it. Men sent from the gods are taking our soldiers. They must need them to fight in their world. By Jupiter and Vesta, I don't know why they need them. They are replacing them with men not of this earth, not real; but look and move real. Why do they do this? Why do they do this!"

Dr. Brown and I were both fascinated and confused by this. We both looked at each other and shrugged. Then I saw a spark in Dr. Brown's eyes. His flat expression transformed into a cheshire cat grin. "They're taking soldiers," he exclaimed. "The government is taking roman soldiers!"

Chapter 20

Dr. Brown and I were discussing our next move in a long university hallway packed with students. I stared earnestly at him, while thinking about our next set of actions.

"So, what do we do now," I exhorted.

"Let's go to the press, the media would love this story--right?"

"No, the government controls the news media, trust me on this one. I can't go into details, but I know this for certain."

"Well, there's always the internet. We could create our own anonymous blog that reveals this truth to people!"

"Yes, we'll do that. But there's something else we should do; let me think...I know, I'll write a book about my experiences! We'll publish it ourselves, so the government can't get to the publishing houses and alter it in any way."

"Ok then. This would make a great book! Time travel, history, a revelation of what American has become. Who wouldn't read all about it?"

"Yep, but somehow I think the best is yet to come!"

"I think we need to go back to England and rev our time machine up."

"Yeah, and see if we can get in touch with that Lou Parn guy again. He was helpful."

"Sounds good."

We traveled back to Carlisle, and rented the same flat we previously used. We rented the old barn we were using to store the time machine under the University name. To advertise the fact we were coming, we put our names and addresses in the local phone directory. We hoped if this Lou Parn was as clever as we thought he was, he'd find out we were coming back. This did carry some risks. The government would know we were coming as well. But it was a risk we were willing to take in order to gather more information from this man.

After about a week after our arrival, there was no sign of Mr. Parn at all. No letters, clues, or hints of his whereabouts. Then on the tenth day we were back in England, as Dr. Brown and I were walking to our flat from a local convenience store; a black Chevy Caprice pulls over towards us. Tinted windows, dark exterior, the car screamed stealth. The driver's passenger window electronically rolled down to reveal a 9MM barrel pointed to our faces as we hear the fateful words,"Get In!"

We leap into the car's backseat nervously anticipating being kidnapped, charged, or even murdered. He whisked away, driving quickly--but flowing with traffic.

The driver had a rather soothing if raspy voice and said, "I'm Lou Parn. It was stupid for you to come back and make you're arrival so obvious. Did you want to find me, or did you want me to find you?"

"We wanted to find you. You're a valuable resource to us. We wanted more information," Dr. Brown said.

"We knew this was risky, but we were willing to take the risk," I said.

"It was risky, you should've stayed in America," Mr. Parn stated.

"May we ask a few more questions," Dr. Brown queried.

"Alright, but only a few and I may not answer all of them," the mysterious stranger said.

"How long has the government been taking roman soldiers to put in their own military?" Dr. Brown asked.

"They just started recently, this is just the beginning of this program. The goal is to raise 10 million troops, from different periods in history," the driver answered.

"My God, that's a force that could conquer any country on earth!" I screamed.

"Yup, and that's just the beginning, with time travel anything's possible. It's certainly the world's greatest weapon. Of course we're going to take english soldiers from the British empire. No need to teach them a new language. Easier to integrate into the US military. I'm going to drop you guys back to your flat. But go home! I have ways to get in touch with you guys. If something important happens, I'll notify you."

With that he dropped us off at the flat. We soon booked a flight back home satisfied with the new information we just received. More chapters and more revelations for our book. I wondered who Mr. Parn really was, and why he wanted to help us. I knew however he was a man we could trust. I mean...we were both still alive.

Chapter 21

"Look back over the past with its changing empires that rose and fell and you can forsee the changing future, too." Marcus Aurelius, emperor of Rome.

Creedo was speaking to those in the government who controlled the media and entertainment in the United States. It was back in the Pentagon in a large rectangular room with a conference table that could seat about 40. Entertainment interested Creedo. It was quite popular to divert the masses in Roman times and he perceived parallels here.

"So as you know we Romans had plenty of diversions to entertain our people. It helped promote our control over our citizens. I've seen your invention called television, but I haven't watched it in a while. What programs do you put on these days?"

"Oh yes, we have diversions too. Currently we are dumbing down the American public with endless "reality" shows, which are basically scripted shows that people think are based on real peoples lives and situations." One of the programmers responded.

"I have a suggestion. Why not put gladiator fighting on the television. Your masses would enjoy it thoroughly. Our people did. I think the Americans would too!"

"You know...it's funny. We're doing that. It's quite a popular program. It stars a former wrestler and a daughter of a former boxer. It's a bit of reality fluff to divert the masses. Like the Romans did."

"You're joking right? After 2000 years your people revived gladiator fights?!"

"Well kind of. They fight with pillows and maneuver around obstacle courses. No one gets killed. It's quite a popular show. We're also putting extreme fighting in prime time. We were inspired by the romans to do this. It's quite effective at diverting the masses."

"I have a suggestion. Make it real! If your society is progressing like I think it is, your people want real death, real mortal combat, and real blood. Revive real, authentic gladitorial combat just like in the Roman times!"

"We can't quite do that. We have ethical standards these days. Although it would be great for ratings, profits, and market share."

"If the Americans favor any kind of gladiator fighting, they thirst for real blood. Make it real and it would be the most popular entertainment you ever saw!"

"Hey, if we could we would. Real gladiator fighting huh? Maybe it could be voluntary. Or make those Gitmo prisoners fight. They basically have no rights anyway. We might seriously give this some thought. Let me run it by my superiors."

"This is my suggestion. Are you really serious that a gladiator show is on your television?"

"Absolutely. It's quite popular."

"Amazing. Will the similarities never end...I mean really?"

"Probably not."

This conversation really inspired Creedo. He imagined reviving something he thought long gone. The games as they were known, enthralled and excited the masses like nothing else. Men fought to the death in mortal combat. Some were devoured by lions. They even did mock naval battles by flooding the amphitheatre with water from aqueducts in Rome. The thought then rushed to his head. This new empire must rebuild their own coliseum exactly like the Roman one. But bigger, better, and even more violent. He would discuss these proposals with those who controlled both the media and military-industrial complex in the United States. The corporate elite. Similar in power and influence to the Roman plutocrats in his time, Creedo knew who was pulling the strings of government here. He also was extremely clever and adept and manipulating them. So if the games had to be revived in all their former glory, a new amphitheater had to be built. He had some convincing to do.

Chapter 22

Dr. Brown and I reviewed our notes. We contemplated our next plan of action. I started to write my book. I thought about contacting Creedo. Although we had a falling out, I thought I could at least get some information from him. So I went to his office in the university where he taught to grill him for information. As I walk in Creedo is sitting at his desk, grading exams for one of his classes. As I open the door he cocks his head and glares with piercing eyes. I stammered:

"Hi Creedo, how have you been."

"How are you Mike. Sorry we haven't been on the best of terms lately. Can I get you something to drink?"

"Hey that's okay Creedo. How's your classes been?"

"Going well. The students love me. I explain how Roman life was really like.

They seem to sense I'm more authentic than some other teachers. They still don't know my real past, but I tell them what I can about life in the ancient world. How are you doing?"

"Some interesting stuff has come up. I wanted to go over it with you. I've been going back and forth through time and I seemed to have discovered some stuff the government is doing. Some really strange things. Like impressing soldiers from your era. I don't know what to make of it really."

"Well. What can you do about it anyway right. We're pretty insignificant really compared to the empires we live in."

"Yes that's right. But there is one thing a little person like me can do to change the world."

"Oh really? What could that possibly be?"

"To write a book."

"What good could that possibly do?"

"You'd be surprised. Great books have changed the world. Thomas Paine's Common Sense helped fuel the American Revolution, Harriet Breecher Stowe's Uncle Tom's Cabin assisted the cause for the Civil War. In fact Abraham Lincoln said she was the little lady who started the Civil War. Karl Marx's books started the Communist revolution. You'd be surprised what great books can do."

"Yes, but I think your people are deliberately dumbed down to the point where they don't even read books anymore. Distracted by bread and circuses. I don't see what one book is going to do."

"I'm going to write on anyway. So any ideas why the government is stealing troops from the past?"

"To use in the present."

"What for? I mean the logistics would be tremendously difficult. They'd have to learn a new language, new culture, new weaponry. Wouldn't that be more trouble than it's worth?"

"Not if they didn't have enough troops here."

"Well that's true. And with time travel they could always have enough time to train them for combat. Well I'm going to lay low for a while. You probably won't be able to contact me again for a few months."

"Where will you be?"

"Someplace the government can't find me. Like the mountains of Waziristan. Hahah. Anyway best of luck to you Creedo. I'm glad we had this little chat."

"Yes, me too. Good luck."

Dr. Brown and I were indeed laying low for a while. We stayed in various cities, in more or less disguises, as transients. Meanwhile I was taking notes and outlining my book about my adventures. The government's secret program to amass a huge army to build a vast empire was still underway. What could the two of us, or this Lou Parn fellow, do about it anyway. I was starting to come to the conclusion that Creedo was right. Just let events fall as they may. Que Cera Cera.

Chapter 23

The president was in the cabinet room discussing policy. In the room as an invited and special guest was Creedo; whom the president took a liking to. He convened the meeting to discuss the Supreme Court's decision to strike down the Military Commission Act which would have expanded the president's powers to suspend habeus corpus for suspected terrorists. The president wanted to discuss ways to deal with the decision.

"Well, the bastards took this power away. I don't think it's good for national security," the president said.

"You know when I said this power made you basically an emperor. Well it doesn't. You still can go to war without a declaration from Congress, you still have the immense powers of the Patriot Act to search and seize peoples records without a warrant, the CIA can always rendition people to other countries to interrogate them. The NSA can wiretap citizens. I could go on and on. I advise to just have the CIA rendition these prisoners to countries that can really take care of them," Creedo advised.

"That's a good idea. But those guys really piss me off. How dare they endanger national security like that. I just wish I could get 'em back for this!" the leader exclaimed.

"If that's your wish, I have an idea you might like," Creedo put forward.

"What is it Creedo?" the president wondered.

"I'll tell you in private, I think you and a many other citizens will enjoy it," Creedo answered.

"Tell me Creedo, I'd like to hear it," the president demanded.

"In due time sir, with all due respect in due time," Creedo insisted.

"Well anyway renditioning these terrorists is a good idea. They don't even need a trial or anything. You know what Syria and Egypt do to these guys. They'll wish they were stuck in Gitmo!" the president stated.

"And thoses judges will wish they never made you angry, I'll tell you that much," said Creedo.

"I'm looking forward to your suggestion," the president concluded.

A strange yet professional voice I didn't recognize called me on my private cell phone. He sounded nervous yet reassuring. He talked quickly and in an austere manner. We were hiding out in a dumpy little motel outside Chicago.

"I must talk to you," the strange voice stated.

"Who is this?! This isn't Lou Parn, it doesn't sound like you."

"No, but he's a good friend of mine. You can trust him."

"Ok, but can I trust you. Who the hell is this and how did you get my number?"

"We've got all the numbers. And we can track anyone. Never mind that. I'll meet you at your motel room tomorrow at 8 AM. I'll give you a package. Be there."

"Another mysterious stranger with a package. Ok...well it was nice talking to you 'mysterious stranger with a package.' "

Jim Douglas was working at the NSA field office in Chicago. He worked long hours and put in more than a fair day's work. He was staying at a cousin's house while in Chicago, for he was just working there temporarily on a special assignment. He got up at 6:45 AM to make the package run. He took a cab about halfway from Oak Ridge to Chicago. Then he assigned another cab to a rental car service, and with an assumed name rented a car to his final destination. He wore a blonde wig, sea blue contacts, nondescript Chicago Bulls T-Shirt, jeans, and white tennis shoes. He really looked like anyone, and that was his plan. When he arrived at the motel desk, he told the clerk to he wanted to see Room 115. "They're expecting you," he said.

Douglas turned his back away from the desk clerk. The clerk lurched down towards his left leg and grabbed for something. He quickly grasped a Smith and Wesson 9 mm semi-automatic pistol that he fired at Douglas's head. Douglas who had a sixth sense about this kind of situation immediately swirved while ducking and drew his much larger Browning 9mm Hi-Power GP pistol. The clerk fired his gun a split second too late. Douglas fired his round and made a direct hit at the clerk's forehead. The commotion caused a stir at the motel. Douglas ran towards the motel room, knocked on the door and yelled, "Here!" Then darted off.

Startled I grabbed the package and hid it between the mattress of my motel bed. I heard the gunshots and ran to see what the commotion was about. Dr. Brown soon followed. We fled the motel before a fleet of police, EMT, and rescue personnel came to the scene. We grabbed the package and our immediate belongings and left the Chicago area for parts unknown.

Chapter 24

The grand construction project will be underway. The RFK Stadium in Washington DC is slated to be torn down for something more awe-inspiring and historic in nature. The project underway would entail something not seen in almost 2000 years. The United States government will reconstruct and double in size the original Flavian Amphitheater or Colosseum in Rome. It would be over 300 feet tall and 1200 feet long (almost doubling the dimensions of the original). With a base of 12 acres the size of this amphitheatre would be gigantic even by today's standards. A perimeter of more than 3600 feet, it will be made of marble and concrete and travertine stone just like the original. It will also contain a hypogeum under the floor and a piping system allowing it to flood for naval reenactments and water shows. It will also be canopied by a velarium, a cloth covering to shield the sun from spectators. An exact replica of the original it will also give the public what it wants--blood. All this of course, takes place well into the future.

Douglas couldn't figure out how the clerk at the motel knew he would be there. He covered his ass. He told no one about his stint in Chicago or what he was planning to do. He was careful. The clerk knew he would be there, knew agent Douglas's description, knew the target. He might even had known his mission. He didn't know how he was compromised. He had his methods of finding out where I was and Professor Brown, but how did the poser at the counter know all this? The only person he had told was his most trusted and loyal partner--Agent Fields.

He knew it couldn't be her. Why would she betray him? They had been partners for years and he knew he could trust her. She told him everything; and vise versa. But if not her, who else could have known about this mission, these contacts, and this location. He thought long and hard while sitting at his field office in Chicago. He couldn't come up with anything else. He couldn't come up with anything else. What he could believe however was that someone else on the inside betrayed him. Perhaps planted a GPS device on him or his car, so they'd know his whereabouts. His mind was racing, but couldn't definitively come up with a damn thing.

After we fled Chicago we meandered our way to our lab base in Sedona. We had colleagues there who could help us. We knew the lab itself would be monitored by the feds. But we knew people in the area who--if taken the right precautions--could help us. We knew we couldn't call them, email them, snail mail them or anything like that. So the only way to contact them was (in a disguise) show up at their door.

We were exhausted. Traveling really took its toll on us. When Dr. Glenn Reed opened the door of his one-storey spanish style ranch house he seemed startled, but then his face relaxed--like all the tension fell from his body.

"Can we stay with you for a while," I asked.

"Of course you guys can. You look exhausted. Where have you been and what have you been up to?" he queried.

"That's a long story, and we'll be glad to tell you about it. Let's just say we don't want any imperial entanglements," I stated.

"Oh is Darth Vader after you," he said jokingly.

"Pretty close," Dr. Brown shot back.

We entered his home and sat down in his living room. Dr. Reed headed the particle physics department of the time displacement project. He was our most trusted and vital team member on the project. We filled him in on the details involving our journey. His wife was at work; so it was just me, Dr. Brown, and Dr. Reed. He was very involved and helpful. He had an extra bedroom in the anterior, that he used when his son and his son's wife came to visit. He said we could stay there as long as we needed to. It was a nice house in Sedona, with vast desert mountain vistas in the open Arizona sky visible from the back yard. It was no wonder Sedona has been called America's most beautiful place by many prominent travel magazines. We were surprised and elated that the feds pursuing us didn't deter him from letting us stay there. It was a hospitality that we were eternally grateful for. After graciously serving us dinner consisting of warmed up sweet and sour pork with fried rice, he started asking us a few questions.

"So may I ask why the feds are after you guys?" Dr. Reed wondered.

"Sure, basically we know too much," I answered curtly.

"Yah but you two are the most law abiding citizens I know, why are you in such trouble?" he persisted.

"It involves our time displacement unit. What they're doing with it and what their future plans are. We'll fill you in on the details later. It has to be a very safe and secure environment though," Dr. Brown shot back.

"Ok, well me casa su casa," Dr. Reed offered.

"Thank you so much for your hospitality, we've been on the run for a while. Basically what we know holds the future of the world in the balance. Dr. Brown and I believe no country, not even the United States should have too much power over the world, and the feds are pursuing us for what we know, and the fact that we could just possibly stop them. But we do have help," I stated.

"Good luck to you guys. I know I can believe in you guys," Dr. Reed granted.

"We won't stay more than a few days. We really don't want to put you or your wife in any harm. We just need some time to get to our time displacement unit, then we'll figure out to do from there," I told him.

"I want to come with you guys. You could use some more help, right? I can't pass up an adventure like this," Dr. Reed asked.

"You've been nice enough already. But hey, you know the risks. Talk it over with the wife, okay?" I said.

"No way. If it was up to her I'd have no life. I'm coming with you. There's no stopping me!" he insisted.

"Okay, it's off to the lab tomorrow. Tomorrow's going to be an eventful day," I declared.

"Yes, I can't wait to see what the future holds for us," Dr. Brown concluded.

Chapter 25

Douglas was in his Ft. Meade, Maryland office contemplating what went wrong in Chicago. He had to talk with Agent Fields just to put his mind to rest. He was working on the massive ECHELON program at this time, basically an international project consisting of the US, UK, Australia, and New Zealand to monitor commercal satellite signals circling the earth. But his mind was on Stacy, and whether or not she possibly could have betrayed him. So he sent one of his clerks to personally ask her to meet with him at a local coffee shop the next day. He never used his cell or any other electronic device to contact a NSA agent. He knew all these forms of communications were monitored at all times. They met for espressos the following morning. It was a bright morning, around 8 a.m. she came in with a dark suit looking professional as always, her eyes looking colder than usual.

"Agent Fields, nice to see you again," he engaged.

"You too Douglas, what's up?" she inquired.

"You heard about what happened in Chicago. Just trying to get to the bottom of it. Trying to find the mole. I know there's somebody out to get me at Mead, just don't have any leads yet," agent Douglas said.

"Yes, I heard about that. You could've been killed! I'm glad you're ok. I think there's a mole too. Better be on your toes. I have no idea who it could be. Just better be careful!" she volunteered.

"I will, I will...I just think that the mole is somebody we know. Somebody who was on the team; the team in England. What do you think?" he asked.

"That sounds plausible. There were a bunch of guys who I never met before. I can't speak for them one way or another. You could look into them." she suggested.

"I'll do that. Hey this is good coffee. Need the caffeine rush in the morning. Hey...by the way, where did you hear about the incident in Chicago?" Agent Douglas wondered.

"I was in my office here at Meade. I was devastated to hear the news. I'm so glad you were'nt hurt. Usually the cryptography business isn't this dangerous," she offered.

"Oh, we both know that isn't true. All intelligence work is dangerous. Don't let the office work and bureaucracy lull you into a false sense of security. We need to be on our guard at all times. At all times," he concluded.

Creedo was in the Eisenhower building at a spacious conference table with his new compatriots. Speaking with the vice-president about matters of state, he observed remarkable similarites between our two political parties, and the optimates and populares parties of his day. He was both impressed and intrigued by what was revealed.

"So, Mr. Vice-President. Your political entities the Democratic and Republicans are very similar to the optimates and the populares of our day. The opimates are like the Republicans, who favor the plutocrats and the wealthy; and the Democrats are like the populares party of which Julius Caesar was a member. Was this set of events taken directly from the Roman's?" Creedo queried.

"Oh that's just what people here tend to believe. The truth is both parties merged years ago. We have an agreement to share power. It's like a marriage, a bickering one sometimes, but a marriage nonetheless. In fact, it couldn't be represented better than an actual marriage between two of our best campaign advisors. The campaigns and elections are just for show," the vice-president revealed.

"That's brilliant. Our parties were so sharply opposed to each other they formed private armies and fought wars. Yours merged into a single unit. This must create peace in your nation!" he declared.

"Yah, the points we differ on are trivial and media distractions. We agree on the fundamentals like national security, drug policy, monetary policy, foreign policy, overseas military commitments, intelligence matters. There's basically one political organization running this country. We just allow people to think there's two of them so people believe there's democracy here," the VP gloated.

"I really respect that. So the country is really no longer a republic?" Creedo asked in amazement.

"It is to us. Hahahaha," he chuckled.

"Yah, just the other day, I was speaking to a citizen who said that no matter which party or who gets into office, nothing seems to change. Well in actuality, he's absolutely right. So there's basically one organization that rules this country," he expanded.

"Well. How did that happen?" Creedo inquired.

"Money. Both parties get their funding from the same sources. The same companies, and in many cases the same individuals. We realized that there's no real difference between our organizations, so we secretly merged our operations. It's probably the greatest scam in American history; next to Social Security. But that's another topic," The vice-president emphasized.

"Very good. That's fine with me. However you people run your society is acceptable. You may avoid the kinds of class and political warfare we had. Your country is better off with this arrangement," Creedo concluded.

"We think so. Hey look I gotta run. Affairs of state and all that," The VP finished.

"Another great discussion with the rulers of state. I wish you well co-imperator!" Creedo finished.

Chapter 26

We entered the Sedona lab early in the morning. We knew of a secret annex that no one else besides us knew about. We revealed this to Dr. Reed to show him the prototype time machine that we hoped the government could not hide or steal. We were going to manipulate time again to determine what the government was going to do. This was do or die. I went inside and attempted to boot it up. To my great displeasure it would not work. Dr. Brown and I attempted to fix wires and machinery, but it still wouldn't operate. It needed serious work, and we did not have the parts on hand in the lab to fix it. We would have to stay with Dr. Reed a little while longer, so we could get the parts needed to repair the unit. I was pissed, but there was nothing more we could do. At least Dr. Reed had connections to get the parts; in the meantime we could plan to reveal the truth to the American people--about what was going on.

"We failed to get him," the agent exclaimed.

"I know--we certainly tried." the conspirator answered.

"How did you know he would be at that motel?" the agent queried.

"A fortunate slip-up by a Dr. Harold Brown. Well, on second thought maybe not so fortunate. He used a credit card to purchase some items from a convenience store right next door. From that we determined where they were staying. We sent a van out to track and observe them. Then we detained the owner and staff of the motel, sent in our own guy, and had him pose as a motel clerk. Next time we won't use a rookie--big mistake," the conspirator observed.

"Yep, we knew Douglas was a loose cannon for a while now. He's basically committing treason by helping a couple of rogue scientists who want to subvert our mission. We need to take him down. Arresting him is impossible. It would jeopardize everything on this project. As if we could present evidence on him or anything. All we can do is neutralize this threat. So...what do you think Agent Fields?"

"I'll take care of that soon. Trust me!" she exclaimed.

The time displacement unit was almost operational again. We got the parts and fixed what was wrong, but had to wait to get funding for the power source. We got the funding through our respective universities. We told them it was for a top-secret project involving new sources of energy. The plutonium power source was then online, so we were ready to go. But go where? We still had to agree on when, where, and why to travel through time again.

When we arrived at the site of the time displacement unit, Dr. Brown sensed something was wrong. The chamber where the displacement unit was held, seemed a bit different than before. We searched around, then from behind me I heard a ruffling. Then when I immediately turned around I saw a man holding a gun on us. When I saw his bloated eyes staring at us he exclaimed, "I"m Agent Douglas from the NSA. But don't worry, I'm on your side."

"I'm Harold Brown. We met before, what do you want?"

"Just to talk, fill you guys in on some stuff. They want to kill me so I'm hiding out, like you. I'd like to come with you guys if that's ok."

"How do you know we can trust you? Why do they want to kill you?"

"For giving you guys information. I'll give you my weapons. I'll also tell you how not to be found. I suppose you want to travel through time again?"

"Give us your guns first. We are armed too. There's a couple of guys coming."

"Ok. I'll do it," he hands a couple of revolvers to Dr. Brown.

"Thanks. And thanks for the package, that was helpful. I guess you're coming with us. We need to keep an eye on you. You seem to be alone, if you are then I guess you are telling the truth."

"I am believe me. I hope we established some kind of trust."

"About as much as any NSA agent can establish."

"So...where are we going?"

"We don't know yet. But it'll be an adventure! They always are."

"You can say that again," I replied.

Chapter 27

Before traveling through time again, we invited Agent Douglas to Dr. Reed's house. We wanted to get a sense of whether we could trust him. We had a few questions for him before we could activate the time displacement unit again--and take him with us. He spoke with confidence and force, like a typical federal agent, but I could sense a touch of melancholy to his voice about the betrayal he felt, especially regarding his partner. They had been close, tight partners, and he seemed to be certain that now she was trying to kill him. We spoke for about 20 minutes. Dr. Reed offered him something to drink, before we went back to the Sedona lab to decide when to travel to. When we arrived back at the lab, we decided to travel to the future. We figured if we did that we could determine what the government would do with their time travel program; then after seeing what they would do, we would go back in time again to the present, to figure out whether we should stop them and how. The exciting part about time travel is based on something called string theory and dimensional reality. Since the future contains different dimensions (based on different outcomes of time-space) one can indeed go back to the past and shoot one's father and still be alive in the present. The reason is that a different reality exists in that world, so an action in a different dimension does not affect the current one. In that respect, one can travel to the future see one possible outcome, travel back to the present, and change that future outcome. It is all very complicated and involves quantum physics, string theory, and time-space dimensions; but one needs to be aware of it when traveling through time.

So onto the future it was. Or will be...it is hard to say properly. In any case, we decided 30 years in the future would be adequate to determine the government's intentions. The party would now include myself, Dr. Brown, Dr. Reed, and our new companion Agent Douglas. This was a far larger group than I could ever have imagined. We entered the displacement unit eager, anxious and ready. A sharp jolt-like turbulence shook all of us. It felt like an earthquake at least 6 on the Richter scale. I seriously thought the unit might explode this time, killing all of us inside. But for the grace of God we made it. When the process was complete, we all exited in a rushed but polite manner. Dr. Brown was the one to go first, then myself, then Dr. Reed, then Agent Douglas. "We made it!" I exclaimed. "We made it alive!" Dr. Brown replied.

We now left the facility. I was impressed to view that most of the cars were indeed electric or hydrogen powered. Better for the earth, I thought. I even saw some aerodynamic flying cars. Finally, we all said. They have flying cars in the future. I came across a tall man about 6'2", dirty blonde hair, on the sidewalk and had a few questions for him.

"Hi, do you mind if I ask you a few questions?" I asked.

"No, not at all. Go right ahead," he replied.

"What changes have occurred in the past 30 years?"

"Wow, where you been man, living in a rock. What do you mean?"

"Well, lets start with politics and the world situation."

"Why? You been in a coma or something."

"Yah, and I just came out of it. So fill me in please."

"Well ok. Well, about 30 years ago, the terrorists exploded a nuclear bomb that blew up New York City. The government then declared martial law. They formed something called the American Union so they could go after terrorists in every country. Oh and then...hey remember Canada?"

"Yes, of course."

"Well, it's not there anymore. It was annexed by the American Union. This is the new name for the United States, by the way. Anyway, our president said in order for us to effectively fight terror and Islamic extremism, we needed to take over practically every country to stop them. We annexed Canada, then Mexico. The American Union merged with the European Union. Then they just became the American Union-European Division. Then, get this England became a state. How bout that for irony huh. Then the American Union declared war on Iran, Syria and Pakistan which was taken over by extremists. The whole middle east became part of the American Union. Then we annexed Africa. The president's father was from there, so that made it easier for them. Then with a few hundred thousand troops they conquered South America. They had to take out a few hostile countries there. You know, honestly I don't know where they got all the troops for this. I thought years ago troop strength was wearing thin, but who knows. They don't bring this up on the news media now cuz the government controls it."

"Are you joking! Is all this really true?"

"I'm not kidding man. Look it up on the holonet," he then presented what appeared to be a cellphone that projected a holographic image. It presented a moving image about six feet from me onto the sidewalk. The image was about 3 feet suspended into the air. It presented a brief documentary of everything he told me. I stood there astounded and perplexed at both the technology, and the content. I thanked the man for revealing that information and gathered my group for an impromptu meeting to determine our next set of actions.

"So, what do we do now guys?" I asked.

"Well, since this is the future, I suggest finding a place to stay, a way to get money. Will our money even work here? It seems to be a radically different society from what he told us." Dr. Brown offered.

"Yeah, let's get our bearings. Let's get a video log of this society, bring it back to our own time and let people know what will happen. Let's see if we can get a hotel room for a while. Let's spend at least a few weeks here to make any determinations." I suggested.

"Sounds like a plan. Let's find some place to stay first." Dr. Brown concluded. And with that we all search for lodging for a while.

We walked down the main street in future Sedona for a while and found a Motel 6. It still preserved the name of that establishment. I paused for a moment and stopped the group's stride. "Hey, what if he asks for ID's. They're 30 years old, they'll never work. What do we do then?" Agent Douglas then suggested he flash his badge and tell the clerk we needed a place to stay on NSA business. He entered the office and did just that-- securing us lodging for the forseeable future. We entered our rooms and collapsed on the semi-comfortable queen sized beds. We reserved two rooms, Dr. Brown and I had one room, and Agent Douglas and Dr. Reed the other one. The following day secured more good news pertaining to our stay there. Dr. Brown and I went to a local store to buy some food and supplies. After pulling out some money to pay for our purchases, the cashier shrieked. He saw a ten dollar bill on my hand and recommended I sell those quickly. He asked my and Dr. Brown if we had any other currency and we told him we didn't. He gave us directions to a place that converted old currency into new ones. We followed his advice and went to that store. Fortunately, the place was walking distance, so we offered a few hundred dollars for exchange. After the proprietor checked the dates he screamed in surprise.

"Do you know how much these are worth?" he asked.

"No idea, lets just say we've been away for a while, years even. We've been well out of the country. Haven't followed the news even. What's a few hundred dollars worth?" I asked.

"Well since the great inflation and the government currency confiscation act, this money has gone into the black market. The government destroyed most of this currency and printed million and billion dollar notes. You'll probably get about a hundred and fifty billion dollars for this. No questions asked," the proprietor stated.

We were very excited about this prospect. However, our enthusiasm was tempered by the notion that 150 billion dollars was worth about 15 thousand dollars in today's dollars. Still, this would let us stay in the future for a few months to get an assessment of what things would look like, and what the government planned.

We then were able to rent out some pretty swanky motel rooms for a few months. We were also able to accommodate ourselves some nice transportation. The cars in the future were all electric or hydrogen, so no annoying gas stations or fill ups were necessary. We could just plug our cars in an outlet for a few hours--then our cars were set for days. After our logistics were taken care of, I got a new laptop to write my journal. I was also able to ascertain more from our new acquaintance Jim Douglas. He told me about his relationship with Agent Fields, and her apparent betrayal. He spoke of his love for the NSA; why he felt betrayed by them, and why he joined with us. I asked him to recall specific events, conversations, and other details for my journal. He was surprisingly cooperative. After finding out more from this future world, we purchased facial disguises and traveled to Washington.

Chapter 28

Creedo was in a good position. He was in the front row right square in the middle of the amphitheater. There was a crowd of over three hundred thousand screaming fans piercing his ears. This only increased his excitement. He was watching an ancient ritual he was very familiar with. Two men in the arena fighting to their death for the amusement of the crowd. This was not ancient Rome however, this was America. Thirty years into the future.

Creedo enjoyed the blood; but he enjoyed the fight even more. It was the mortal expression that fascinated him. The look of death on their faces. He advised them on the reconstruction of the new amphitheatre. Now he was experiencing the rewards of their efforts. These men were criminals facing the death penalty; so they had an option of getting lethal injection or subjecting themselves to the games. If they survived the governor agreed to commute their sentence and give their families a huge stipend. If they were killed in the arena, the logic goes, they would face death anyway, so why not go with a bang. Even if they died, their families would be well compensated for their participation; and for the ratings and revenue. The producers of the events agreed to give the states where the participants reside part of the revenue. So everyone made out. There was even a completely voluntary aspect, where usually destitute individuals would sign a release and enter a competitiion for a thousand pounds of gold. Not everyone was killed in this competition, but it was usually up to the victor--and the crowds. Creedo was viewing two prisoners fight to the death in this scenario. They were big men--about 6 foot 4. Stocky build, one of them had a shiny bald scalp with a scar on top. One was weilding a mace and a steel net to confront his opponent, the other a foot long lance with a serrated blade. They were in the duel of death. The bald one with the mace took a step back. He had a focused glare that pierced his opponent. He glided forward and flanked his opponent with the net. He then struck a mortal blow with his mace. His opponent's lifeless body was feast for the hungry crowd.

Due to the terrible economy, the many destitute would agree to perform in gladitorial combat for a fee. In death, their families would be compensated. The promise of reward was worth the risks for them. The new America took its toll on the poor, destitute, and the government used that for entertainment as a diversion from their troubles. The state would provide funding for these "games" and the citizens would watch at the amphitheaters, online, on television or on their cell phones. Ancient entertainment made more accessible by technology.

Reviving the games was also an homage to the new Roman infantry that were assisting the new American superstate. This was the kind of entertainment they were used to and to bring it back was enjoyable for the Roman troops, and the American masses. It was a win-win situation for all involved, the government decided. This is exactly the kind of entertainment that a great empire needs to project. Power, violence, honor, and sacrifice. And most importantly entertaining the masses. Creedo knew it would come back, he just knew it.

Chapter 29

It was Fields' last year at the NSA. She served about 35 years with the agency and was ready to retire. She rose through the ranks and became a well-respected operations director, but she felt she'd been there long enough and didn't want the desk job anymore. She always had wondered what happened to Jim Douglas, her partner, her friend and colleague. He literally just vanished. This wasn't that unusual in the spy business. People "disappear" all the time from the CIA, NSA and such agencies all the time. But no trace of him had ever been found. Usually after 10 years or so, the government finds out they were murdered or assumed an identity to escape--but after 30 years NOTHING! This was truly incomprehensible to Fields. She had to know what happened to him. No matter what it took.

She went to the obligatory dinner, got the gold watch, and went home. She then took a flight to Sedona. The last place Douglas was seen alive. She stayed at the best motel in town. She rented a car and went to the scientific lab where the time machines were built. They were abandoned with no signs of life. No displacement units were there. No people, just rusted out deserted hulks of steel and dark, cavernous shells. She wondered if Jim was lost in time. What happened to him? She would try to find people who last saw him and get answers.

Fields had a sixth sense about certain things. She felt the presence of her former parter. In the field she had an uncanny ability to find people who other people thought were permanently lost. She felt this was the time and place for her former partner Jim Douglas. She couldn't put her finger on it, but she knew her instincts. Her first piece of evidence was talking to a man who heard that some people traded old money for new. They had far more old money than most people. He didn't know where they got it, but they must have had connections to get that much. Old money was destroyed or well hidden in this time. Only someone very brazen or oblivious would have that much old currency around. Fields speculated that they were in a hurry to get out of there. She needed to contact the pawn broker to get more information. Once she arrived, he seemed pensive and rude. Hesitant to offer any assistance.

"Can I ask you a few questions?"

"Sure...ask, doesn't mean I have to answer."

"Ok, then have it your way!" She quickly flashed a government badge, hoping this rube wouldn't look too closely.

"Sorry, I'll be more cooperative officer," he submitted.

"Now, I think there was a few people in here exchanging old money, right?"

"Yah, a few weeks ago. It was kinda weird, they had lots of old money, and seemed to be in a hurry."

"Have any idea where they were going?" She knew this was a long shot.

"No idea ma'am, sorry they were very secretive, and in my business I don't ask questions."

"That's fine I understand, is there anything else you can tell me?"

"No, that's pretty much it. I don't think you'll find them around here."

"Thank you for your time."

Washington will change quite a bit from our time. In the future it will be a commercial and economic center as well as a political center for the United States. More skyscrapers and high-rises as a result. The government owned most of the major corporations in America and this led to the expansion of Washington as a financial center. We stayed at a nondescript motel downtown. We noticed the new stadium that replaced RFK stadium that looked much like the Roman Colosseum. We didn't have time to check it out, but we vowed to get to seeing it eventually. We just started with the mundane details of getting a sense of what the future would hold. Starting off with magazines, newspapers, online information and television would be a start. We got an overview of the changes in technology and events in the past 30 years. Some of us also wanted to look up some old friends.

Chapter 30

"Why did she betray me?! She was my partner and closest confidant," Douglas mused.

"I guess her deepest loyalty was to the government. And yours was to what you thought was right. She just took a divergent path than you did," I replied.

"You know I guess you could look her up. She's probably still alive somewhere," Brown offered.

"No that's too risky. I just want you guys to get your time machine back and use these machines for peaceful, scientific purposes. I can't waste time with Fields."

"But how do we get the prototype time machine away from the government. That's impossible!" I stammered.

"That's just it, the prototype is how they were able to duplicate and make their own time travel units. If we stop them from getting the prototype, we win!" Brown suggested.

"Ok, but for now we'll stay in the future. They won't be looking for us in the future...meaning now of course," I explained.

"And what is that Creedo up to? Let's find out about him if we have time," Dr. Brown suggested.

"We have all the time in the world," I concluded.

Creedo was musing on his second-storey patio of his penthouse in his Hay-Adams suite. A top ranking general knocked on the door. Lazily he went to answer it, the military official strutted in.

"I take it you enjoyed the games; I bet it reminded you of old times," the general said.

"I loved the games. I am so happy you revived them. I knew they would be popular here," said Creedo.

"Great. Well we have plans for you. I'm here to notify you of a position we are offering. Take your time to decide if you'll accept."

"I most likely will accept any offer made by this great country!"

"Well, Europe is basically an American province now. We're placing the capital in Rome. And since with your experience we're offering you the position of governor. It will be available in a few months. Glad to see you're so receptive to the offer."

"Of course general. I'll be glad to accept. Back in Rome, I'd be delighted! Tell your president I gladly accept his position."

"Good. Is there anything else you want?"

"Yes. This may seem like an odd request, but, I'd like to participate in the games."

"But whatever for? People get killed in there."

"I want to see if I'm up to the task. As a roman it's something I need to prove."

"Perhaps an exhibition. Nothing where you'd actually get hurt."

"NO! I want this to be real.I need to prove this to myself. Then I'd be worthy to be your governor."

"As you wish Creedo," with that the general snapped around and exited the penthouse briskly. Creedo was pondering. He was old now, but wanted to participate in the games. It was a Roman thing. To be a man, to be a worthy man. He didn't have to do it. But he had to. As a child he remembered watching the games with his father when he was about 10 years old. He asked his father if he could ever participate in the games like the strong, muscular men he so admired. His father told him that he was of the senatorial class and he was above the games. That he would never be as strong or mighty as those men. Creedo wanted to prove him wrong so violently he could taste the gladiator's blood on his palate. He had to participate. He could taste his chance at last.

Chapter 31

Stretched out over my comfortable queen-sized bed in the motel room I watched a little television. They still have television in the future, but citizens there consider it a quaint, old-fashioned form of entertainment. It seemed to be a reality program with a dozen or so contestants. They panned to the Colosseum looking building we saw in Washington where the old RFK stadium had been. It had the usual format of the contestants bickering, whining and not getting along. They also formed the cliques that became a cliche for these types of programs in my time. They all appeared to be muscular men in their 20's and 30's. Well built, I figured it would be some kind of fighting competition, but wasn't sure what kind. After some kind of silly competition involving running through a muddy pit with snakes; the following segment showed one man-- blonde, about 5'11", stocky putting on a gladiator uniform. I laughed. The other one; taller 6'2", brown hair, leaner, did likewise. In front of a roaring audience of at least a hundred thousand people, these combatants wielded blades, shields, tridents, and nets. They stood about 10 yard from each other initially, the crowd was motionless.
Then a giant bell overhead chimed three times. The match was on! The taller man lunged towards the shorter one with his gladius. His opponent had good reaction time, raising his shield to blunt the impact. The shorter one screamed a shrill cackle; hoping to create a diversion. His opponent just stood there with an icy stare. The short one darted toward him and pulled out his sword, just as he removed it from his sheath, the tall one flanked right and stabbed the shorter one in the jugular. The shorter one twitched for a moment, then fell lifeless to the ground. The crowd was deafening. The tall one grinned while spinning around 180 degrees. Then as he was completing the circle he thrust his sword at the man's neck and decapitated him. He presented the trophy for their consumption.
"Oh wonderful, another fake reality show, with fake gladiator fighting. This one has fake deaths even. Must be CGI effects. The head and blood don't even look real," I said sardonically.
"I wonder how they tell who wins if it's all fake like this. I mean reality shows in our day had legitimate contests. I mean it was stupid stuff like eating bugs and stuff, but how do you tell who wins with choreographed fighting and CGI effects. I mean come on! Fake heads? Is the winner chosen by chance or someting," mused Dr. Brown.
"Who cares. I think we have more important things to worry about. Looks like they have their silly diversions here too."
"Yah, I think Creedo would get a kick of this show. Maybe he suggested it?"
"Yah, would bring him back to the good old days. At least he had **REAL** bloody fun with gladiator contests in his day. This is all makeup and special effects."
"Yup. Almost better when they fought with pillows and sticks. At least someone actually won."
"Hey. You know what. Wouldn't it be crazy if this was actually real!" We then all

had a good laugh and changed the channel.

Fields was on the window seat musing on her first class flight to Dulles. She wanted to speak to her ex-partner again and tell him she didn't mean to betray him. She wanted to make things right after all these years. She knew in her gut that Jim would confront her someday. She also knew he came back to this time to tell her something. She didn't know what exactly, but she knew it must be important. This time she would not betray him. This time they would be on the same side. She was too old to have blind obedience to the government anymore. Her loyalty now was to the truth and her partner-- the closest thing to any meaning in her life. She took out one of those flimsy magazines under the food tray in the seat in front of her. She opened it up and read fell asleep to some trivial article about fine dining in Denver. She dreamt she was with Jim on an eavesdropping mission again. She didn't want to wake up.

She arrived at her upscale Georgetown townhouse. Entered the door and figured out what to do with her retirement. No husband, few friends; she had few options in her post-career life. She lay on the sofa watching a soap opera and fell asleep.

Chapter 32

We went to a local diner getting breakfast one morning, focusing our attention on Dr. Reed. Our appreciation for his help was enormous. But we did speculate on why he left his married life to join us on our uncertain and precarious journey. I started the conversation over pancakes and french vanilla coffee.

"So Glenn, why did you join us? You could've stayed home with your wife in your nice Arizona home. You didn't need to join us."

"Have you ever been married?"

"No, can't say that I have."

"Well, I consider this an escape. Besides this is an exciting journey. To see the future and report back to the past. I wouldn't miss this opportunity for the world."

"Yah, I guess this is a one in a lifetime opportunity. But you had such a settled life. This must be kind of a disruption to that. It might even had gotten dangerous. Do you have any regrets?"

"None at all. A settled life is a boring life. I've never had this much excitement and I think thing's will get even more exciting. I can't wait to see what's next!"

"I can't either. Though, I must say I'm a bit more nervous than you. We clearly didn't have authorization from the government to time travel like this. I read in a newspaper of this era that only the government is authorized to time-travel. But that's in this day. Maybe the statute of limitations apply or something."

"Well, no matter. Don't intend to stay here very long anyway."

"Yah, maybe a few more weeks or so, then we're outta here."

We left the diner heading back for the motel. We walked a few blocks and I thought I saw a man in a suit following us. Maybe it was just hypersensitvity but when I turned around he disappeared into the shadows of an alleyway, invisible for the time being. Dr. Reed seemed to notice the man too, but the rest of the group didn't notice anything. Suddenly, a shot exploded. Dr. Reed fell on the sidewalk, blood soaked the concrete, gasping for his final breath. We screamed for help. Douglas darted for the perpetrator but was too late. The authorities came and took the body away to a coroner. We drew a blank on what to do next. I reminded the group that with time-travel we could warn him of this event, or even prevent it altogether by going our separate ways. We would just alter the negative events of the future by the slightest gestures or overtures in the present--or past from our perspective. It was easy because we proved string theory. Every life has many possible outcomes. We could alter time and change this one. But we still needed more information to warn the people of the present of their impending future if they didn't take action to stop it. Before it was too late.

They rushed Dr. Reed to the nearest hospital. We all darted after him. They sent him on a gurney to the morgue for an autopsy. Needless to say we were all mortified by the events. We stood outside the medical examiner's office where none of the authorities could hear us. Obviously, he could not be identified. The medical examiner came out to question us.

"Did you guys know this man," he asked.

"No, but I heard the shot. It's so terrible something like this happened," I lied back

"Did you see the man who shot him?"

"No. And no comment. No offense, but shouldn't the police be asking such questions?"

"I need to know this too. He had no identification on him. Well...I need to get back to work. I'm sure the DA will be contacting you guys later," he concluded.

Dr. Brown, Douglas, and myself walked out of the hospital contemplating our next action. The only question now was to go back to the present immediately, or wait it out a few more weeks like originally intended.

"I say we go back now! Save Reed and change the future!" said Brown.

"No, he'd want us to stay. We can always go back later. Look, I have a feeling something big is about to happen. Something in my gut tell me this. I know it sounds crazy but we have to stay. Just a couple more weeks. I promise," I stated.

"I feel it too. I have some things I want to resolve too. We can go back and save him later. We can just warn him or something," Douglas added.

"We don't even need to warn him. We can just go back and do something the slightest bit different. Sneeze, nod our heads simultaneously, wink. The slightest moves alters the future in different dimensions. String theory, which is now string law proves this. We proved the existence of parallel universes. All we have to do is go back 30 years and go home. End of story. Dr. Reed lives happily with his wife."

"I guess not that happily," Interjected Brown.

"Well that's his business. But the important thing is he lives. But before we go back I want to have a journal of what the future holds. This is important for everyone," I looked firmly at the group as I said this to them.

"I hope you guys know what you're doing. We can't go back if they kill us all!" Brown snarled as we walked to our motel.

Chapter 33

We all returned to our respective motel rooms feeling despondent. One of our colleagues was dead. Dead in the future, but alive in the present. We could alter this course of events, but we still had some work to do and information to gather here in the future. Glenn would have wanted it this way. So we all told ourselves.

Douglas entered his room quickly. He lay his tan suede jacked on the bed and went to the bar for a cool glass of whisky. When he turned on the lights he saw a somewhat older but very familiar face. She sat on one of the plush pleather chairs staring with her piercing sea blue eyes. No look of animosity on her face, she just seemed relieved this part of her life could finally be resolved.

"I knew you'd show up," he proclaimed.

"Good to see you Jim. How have you been?" Fields responded.

"How'd you get in here?"

"You know I have my ways! NSA Universal Motel Key. You were easy to find Jim. This was your favorite motel in Washington. And come on, Doug James on your billing statement. Too easy. You were asking me to find you."

"Yes. I know. So I have a question for you."

"I think I know what it is."

"Why did you try to kill me?"

"It was nothing personal Jim. I was serving my country. Honestly I'm glad you survived."

"So how'd you do it? How'd you know where'd I'd be?"

"We tracked you Jim. GPS unit. I planted it on your gun. I know you always have it on you."

"You gave me that gun. You told me a Browning would never let me down. I suspected as much so I changed weapons. A 38 suits me fine now. So what do you want now?"

"Just to make peace with you. Don't worry about me, I'm retired now. I just wanted to say goodbye. And what it meant to call you my partner and my friend."

"You stopped being that when you betrayed me. I don't care whose orders you were following. How could you do that?!"

"Can you ever forgive me Jim. That's all I want to know. Can you ever forgive me."

Jim pulls out his pistol and aims that barrel at her head, "Just go! This conversation never happened. Please just go!" She leaves the room and hotel. He sits there with a blank look on his face on the bed for a few moments.

We met for dinner at a crowded sit-in chicken place not far from the mall. We discussed our future plans, and the need to expedite things considering the current tragedy that occurred. After ordering we sat down at a booth, ready to execute our next moves.

"So, we need to wrap this up. The question is how long are we going to stay, and what details of this time period do we still need to identify," Dr. Brown queried.

"I'd say a couple of weeks... and I've been thinking. I'd really like to talk to Creedo and get his opinions on what's happened," I posited.

"Hell with that guy! Hes an asshole," Brown shot back.

"That Roman guy. He would have some interesting opinions," Douglas added.

"Well here's my opinion. We have a discussion with Creedo, stay exactly two more weeks, then go home. If we can't contact Creedo, we go home in three. I've got a lot of data already on this time period. I just need one more conversation to wrap up my crucial book. It's really important to me guys."

"Ok, but three more weeks max," said Brown.

"Sounds good to me, then we'll go back to the present and save Reed. We have to alter his future, that should be our main concern," Douglas said.

"Agreed," Brown and I concluded.

Chapter 34

The American superstate was constantly at battle. Foreign invaders, despots and terrorists were the lifeblood of the new republic. They always had to have some "enemy" real or conjured to grow the military-industrial complex into a unstoppable monster with its tentacles into every nation and industry on earth. Having subdued every official state the enemies were groups and individuals deemed enemies of the state and opposing this new world order. Creedo, of course, was loving every minute of it. Relaxing on his patio, sipping some wine from his favorite goblet, he was about to fall asleep when he heard three knocks on the door. He reluctantly arose and opened the door. He found a cardboard package by his feet about a foot long and wide. He opened it and inside he found a white tunic with purple bands around the shoulder area. It looked familiar to him; he proceeded to lift the shirt and spread it by its sleeves and stare at it a few moments. Also in the package was a manilla folder. Inside was a note with the words "I'm back!"

He instantly knew the significance of this shirt. It was the shirt he wore when he first traveled through time from ancient Rome to present day America. It was the tunic he preferred to speak in. It gave him confidence and assurance and he was glad to see it again. He also, of course, knew who sent it. He knew this meant I was trying to contact him. I wasn't absolutely certain he could be trusted, but I wanted to contact him nonetheless. He gave it to me as a present, before trying on more modern shirts to fit in better in his new country. I did this to buy time, letting him know of my intentions, while conjuring up a way we could meet without being arrested by government agents, who Creedo was very associated with at this point in his life. I thought maybe he could first meet Jim Douglas. Then, if the coast was clear, Douglas could contact me on his cell and let me know if it was ok to talk to him.

Back at the motel, we were watching tv and eating pizza in Jim's room which was the largest of the three. We pulled up some well-padded chairs on either side of Jim's bed while he was lying in it. An interesting discussion ensued as to why I didn't use my time travel device to significantly alter history for the better.

"So, with this time machine you guys have in your possession, why don't you go back in time and kill Hitler or something. Or stop Kennedy from being killed?" Douglas inquired.

"That's a good question, and I've been thinking hard about that. Because it's not a good idea to significantly alter history. Say you go back in time and kill Hitler. Well perhaps somebody else leads the Nazis. Say that person is just as evil as Hitler, but not as crazy. This new leader is more effective and say has Germany develop the atomic bomb. Or say this new leader doesn't attack Russia. Russia enters the war on Hitler's side, then the Third Reich wins. Say you stop Kennedy from being assassinated, then there's another Cuban missile crisis that leads to nuclear war with the Soviet Union. Remember

Kennedy almost got us into a nuclear war a year before he was assassinated. You never know what the affects of altering the past are. That's why I took an obscure Roman senator from the past who would not be missed by history, instead of Julius Caesar or a Roman emperor or someone more significant," I replied.

"That's a good answer, so even with the best intentions you might make things worse not better."

"Absolutely. Even considering the different dimensions of time-space altering one might have a spill over effect that affects your dimension. It's better not to mess with it," I stated.

"So what about Creedo then...did you give him a chance to go back. It seems unfair that you took someone from their time and took him to a new time and country," queried Douglas.

"I gave him ample opportunities to go home. He steadfastly refused. This may seem like a strange thing to say, but he seems more at home here than I do. So yah I did offer him to go home, but he wanted to stay here. And now he's some kind of big shot with the government. I hope I didn't affect the time stream too much, but I respected Creedo's decision," I concluded.

With that we had a good evening drinking a few beers, finishing our pizza, and bonding. We were ready to go home; but the wildest adventures were yet to come.

Chapter 35

The following day, I made plans for my last conversation with Creedo before I'd finish my book, and hopefully change the events to come. My plan was to see Creedo alone, while Jim or Harold would be at the time machine. If Creedo set me up, I could immediately call one of them, and he would go back in time and change this future scenario. I would let Creedo know this so he wouldn't try anything. He seemed to be a high-ranking official in this new America, prominent and holding plenty of power. This seemed to suit him, a respected Roman senator in his time, he seemed to achieve a similar stature in this time as well.

Early that morning, I was walking to get my morning cup of joe. A man of average stature, dirty blonde hair, and gaunt features approached me. Seemingly nonthreatening, he approached me and asked me if we could talk and buy me coffee. Since we would be out in the public, I didn't see anything wrong with that and he really didn't seem like the type to be in law enforcement or the government. He chose a booth away from the other customers so we wouldn't be over heard.

"I'm about to reveal something to you," he told me.

"What. And who are you?" I wanted to know.

"I used to work for a great man who's no longer with us. A great man indeed. You knew him as Lou Parn."

"Really?! He helped us out greatly. Who was he and why did he help us?"

"Lou Parn believe it or not was a congressman. The only good one as far as I'm concerned. The only one who leveled with the American people."

"So how did you find me?"

"We've been tracking and following Creedo. We knew you'd try to find him sooner or later. When you delivered that package to him we followed you back to your motel. Don't worry though, we're certainly not with the government. We intend to expose them."

"So what do you want to tell me?"

"Look we better go to a safer location. Hey look I'm not a cop or anything. I'm not armed," he then stoop up emptied his pockets, showed me his socks and seemed clean, "follow me ok." With that I followed him to his apartment a few blocks from the coffee house. When he opened the door a vaultlike room appeared with metallic walls. He told me it was impossible to tap.

"Ok. I will now reveal to you the greatest secret in American history." he said.

"What is that."

"It regards who really runs this country," he solemnly stated.

"Ok who is it? The Masons, Skull and Bones, Bildebergers, Rothchilds, Trilateral commission, all of them?" I asked sarcastically.

"Oh No. No, no, no. The government loves such conspiracy theories. It throws people off. Way off. The truth I find far more interesting. The truth is the policy of permanent government."

"What is that exactly?"

"The fact that all the entities in government you think are competing and separate are united. Democrats, Republicans, conservatives and liberals, the political parties, military, congress they've all merged. The elections are a sham. Just a beauty contest. Notice how they all bicker about the petty issues but all agree on the big ones. Not one told us the truth on the war on drugs, federal reserve, military-industrial complex, or anything important except one lonely man."

"Lou Parn?"

"You got it. He revealed this to people. He wasn't part of the permanent government. He told the truth no other politicians would ever admit to regarding monetary policy, expansion of the empire, government power, and individual liberties."

"Wow. I'm surprised he wasn't killed."

"Well, they did something worse. They marginalized him, and made people think he was a nut. You know what they say about people like him, conspiracy nut and all that. Fact of the matter is he was 100% right and the government we have now is proof of that."

"And you are. I haven't got your name or association with this yet."

"I'm Ed Burnham. I was his Chief of Staff. I take it you used a time machine. You would look much older otherwise."

"You knew about that."

"Yah, and we know how much the military uses it too."

"So what about the Masons don't they have any power or anything?"

"No, I used to belong to them. Their biggest secret is they drink alot and bring over hookers every weekend."

"So why do so many people think they have so much power and influence?"

"So their wives don't know what's really going on I suppose. Look this is real life I'm talking about, not some far-fetched puzzle novel."

"Our founding fathers were prominent Masons doesn't that mean anything?"

"They drank alot and had very active libidos. The Masons were always party animals. Drop it ok?"

"Well look. I'll go back in time to alter this. I'm writing a journal that I hope to turn into a book. I'll make sure to include Lou Parn in it. He sounds like a great man, and he helped me. Hopefully we can get our republic back."

"I think that's a hopeless cause. They can even use the truth against you."

"I know that Ed. I've learned that believe me," I concluded. With that I shook his hand firmly, and walked back to my motel knowing I had an ally in this new time.

Explaining this new character I met to my colleagues would be a difficult task. They obviously would scold me about going with a strange man who could've been with the government, or that he could still be a plant. But there was something I could trust about that man. He spoke sincerely and I could sense he wanted me and my friends to learn the truth. And I was more than glad to receive it. I slept soundly that evening knowing I had taken one step closer to going home.

Chapter 36

"You did what!" Douglas scolded. We were sitting around a small, oval shaped breakfast table eating bagels with cream cheese, after I explained what happened.

"Look, he's with Lou Parn. We can trust me. And he told me some things. I know it sounds stupid to follow a stranger to his apartment. But he's one of the good guys. I know it!" I explained.

"Look, you could've gotten yourself killed. Or arrested. That was dumb. What did he tell you?"

"Basically that Lou Parn was a congressmen. He stood up to the military-industrial complex and the other politicians. That kind of stuff. He served as his chief of staff." I retorted.

"You need to be alot more careful around here. You never know who you can trust." Just then Brown came out to the lobby to join us. He had a small self-contained box of corn flakes and poured some milk into it.

"Guess what this nutjob did last night?" Douglas shouted.

"What's with you two?" Brown was puzzled.

"Mike went off with a total stranger to his apartment who said he knew Lou Parn. Isn't that stupid?"

"I think he knows what he's doing. Lou Parn was a good guy." Brown shot back in my defense.

"He revealed many things to me. Especially about our political system. Not that I can really do anything about it. But I hope the book I'm writing sheds some light on the matter."

"Hey is Lou Parn even his real name?" Asked Douglas.

"No, that's his pseudonym. He didn't say what his real name was and I didn't ask. I looked up all the congressmen from thirty years ago online. Not one by the name of Lou Parn. I think I'll just leave it at that for now." I explained.

"So what do we do now?" Brown questioned.

"I have one last conversation with Creedo, then we can go home!" I said excitedly.

"OK. Well do it then." Douglas and Brown concluded in unison.

Later that day I went over to Ed Burnham's place again. I had an idea. I could talk to Creedo at Ed Burnham's place. A third-party neutral location, it would be a perfect place to have the conversation. I discussed this with Ed as well as some clarifications of what he told me earlier.

"So Ed. You said the political parties have merged. Why was that?" I asked.

"Follow the money. Ask yourself who controls the parties?"

"The national committees, the president?"

"Absolutely wrong. It's the groups that fund the parties. Now a while back there used to be a clear distinction between those who funded the Democrats and Republicans right?"

"Yah, I guess so."

"Such as the big corporations funded the Republicans and the unions, trial lawyers, and liberal activist groups funded the Democrats."

"Ok. so what's your point."

"Just follow the money. Ever since oh about the 1990's, these big donors gave equally to both parties. Hedging their bets so to speak."

"Yes I suppose they did. And that made them merge?"

"They didn't even have to. They became one and the same after that. You can even follow their donations online. The truth is in front of all our faces. That stuff about pentagrams in the capital, funny symbols on the back of the dollar bill, all meaningless diversions compared to this fact."

"So what do you want me to do about it?"

"You could put it in your book."

"Ok Ed. I promise you that. Now there's someone I'd like to bring over to your place soon..." The following morning Ed Burnam used a courier to leave a message at Creedo's door. It said to meet at Ed's place alone. Ed would then search Creedo for anything he could use to contact the government. Once Creedo was cleared Ed would let me know to come over and talk to Creedo one last time in this time period. If that was okay with him, Creedo would leave a yes or no response under a magazine next to his door. His response was "yes" so we proceeded with the interview the following day.

Chapter 37

Ed walked up to Creedo's penthouse the following morning. He knocked on the door in anxious anticipation. Creedo confidently strided out the threshold and proclaimed, "Lets do this." Ed responded, "We'll do this on our terms." He then proceeded to blindfold Creedo and take him in Ed's sedan to his apartment. Before he entered Ed's car, Ed searched for any GPS or any other device Creedo had on him to determine his location. Creedo was cleared. Once they got there, I was anxiously anticipating their arrival. I opened the door when I heard Ed's knock on the door.

"Welcome Creedo, long time no see. How have you been. I hear you've done pretty well for yourself," I opened.

"It's good to see you again too Mike. The times have improved since we last spoke. Things are more honest now. I do like what's happened to your country," Creedo volunteered.

"Well we have our separate opinions on that. It's fine you believe that though. I just wanted one last conversation with you before I left. Wanted to get your observations and opinions on what's happened. See I traveled 30 years in the future so these changes are a shock to me. The new empire is more in line with your philosophy, so I wanted to know what you thought."

"Yes. The United States formally became an empire during that time. It's more efficient and effective now than pretending to still be a republic. I've done well for myself too. I will be the governor of the European province based in Rome. How appropriate is that! America is now the new Rome and can dominate the world under a grand military and administrative regime based on troops from my time! It couldn've gone better for me and your country."

"Well I'd say it's your country more than mine. You've always wanted it this way... and I guess you got it."

"Yes, it was inevitable really. You could've continued being a very inefficiently run republic with vassal states deep in debt, or a glorious empire demanding tribute from your territories. I strongly suggested the latter."

"And you got it. How much did you have to do with how things turned out?"

"Not much. I may have accelerated the process, but the government was going to do this sooner or later."

"I suppose you're right. You look really good for your age. You must be about 80 now?"

"Yes, I'm taking a new anti-aging pill that makes people look and feel at least 30 years younger. It's fantanstic!"

"Nice. Well I won't be staying that long to worry about that myself. It's good to see it works for you though."

"Absolutely. And how have you been Mike?"

"Busy. Moving around a lot. How else have you been Creedo? Have you married or anything?"

"No. I love to host orgies. Also I go to many of what you people call gentlemen clubs. I love those women. They remind me of roman prostitutes. Very nice!"

"Yah that sounds like you. Haven't changed much. So what have you been doing all these years? How did you acquire your new position?"

"They wanted someone who could lead their new Roman soldiers. Who better than a roman leader. See in my time a person of my position had to prove themselves in our legions, we had to rise through the ranks and prove ourselves in battle. I did. The American government appointed me to serve as a liaison between them and the new troops they captured from ancient Rome. I proved myself a capable leader--so I got where I am today."

"Good for you. You know I'm sorry we had to part ways. But we had our disagreements. I'm only going to be in this time for a couple weeks. I just wanted to see how the future would turn out. My time is about 30 years ago, so I will be going back. I honestly wish you well, and like I offered before if you want to go back to ancient Rome to see your wife and kids let me know."

"Never. I'm content to live here and now."

"Ok. Is there anything else you want to know from me?"

"Yes. How is Harold Brown?"

"Good. I hope you're not still holding a grudge?"

"No. Not at all. In fact I'd like to talk to him. Would you mind telling him that?"

"I'll give him the message. What do want to say to him?"

"That's between him and me if you don't mind. I'd just really like to speak with him if that's ok?"

"Fine with me. I'll pass on the message. It was nice to see you again Creedo. This went much better than I thought."

"It was nice to see you again too. I would never betray you. You brought me to this great land."

"Goodbye. And I'll tell Brown you want to see him," with that Burnam put back the blindfold on Creedo, escorted him back into his car, and drove away. I would relay the message back to Brown. Though I was convinced Brown would certainly decline the invitation, being a man of my word I would at least tell him Creedo wanted to speak with him. I didn't see the harm of it anyway, especially since the specter of government agents arresting us had passed. If that would've happened, it would've happened by then, and surprisingly enough--it didn't.

Chapter 38

The aroma of fresh coffee in the kitchen eventually got us to our feet from a long night's slumber. We drank it in morning silence, the period between sleep and living right before a cup's jolt of caffeine tips the balance in living's favor. We all came to my room for coffee and to discuss my conversation with Creedo the previous day.
After slowly sipping my cup away, I volunteered to shatter the silence.

"The conversation went well, very well. I think I made peace with Creedo--and him with me."

"I think it was stupid to meet with him. He works for the government, and the government's after us. How could you risk it?" Dr. Brown responded.

"Ed Burnham made sure it was safe. Blindfolded him and everything. We took all our precautions. He feels a deep obligation to me. I was the one who brought him to this world. He feels he owes me," I shot back, feeling the need to defend myself.

"Well what did he say that was so important?" Brown demanded.

"Basically just some political stuff...one thing he requested though really surprised me," I volunteered.

"What's that?" Brown sniffed.

"Actually it concerns you. He wanted to talk to you for some reason. I promised I'd ask you, but I'm certain of the answer. So I'm telling you now Creedo wants to talk to you. How about that! Well I thought I'd go through the formality anyway."

"Believe it or not I'm up for that. If it's safe to speak to him. I guess you proved that. I suppose after 30 years for him he'd let bygones be bygones."

"I'm surprised to hear you say that. Well if it's ok with you then, I'll arrange it. We'll blindfold him again and make sure it's safe. I suppose you guys have a lot to say."

"And then can we go?" Douglas interjected.

"Actually Jim you should go soon. Today even. Get back to the time machine and I'll keep in contact with you. We should separate ASAP. I'll finish up my journal, but you should get back to the displacement unit and be ready to travel back in time at a moment's notice. Brown and I will stay in Washington, but in different motels. We'll catch up with you soon enough."

"I'll do that. Sounds like a plan. And you need to do whatever else you need to do and I'll see you guys 30 years ago!" Jim exclaimed.

"As far as me Mike. You're not going to believe this, but I'll talk to Creedo. If he wants to talk to me--well I'll talk to him. I'd like to know what he wants to say to me. So tell him I'll see him again if he's still up for it."

"I'm surprised to hear you say that. I'll have it arranged. I'd like to come along if that's ok with both of you."

"It's fine by me."

"And I assume it would be alright with Creedo too. It's on then. This should be a very interesting discussion."

"You bet...you bet," Brown wistfully concluded.

Chapter 39

Creedo reached for his gladius from the sheath on his belt. He protruded it into a hunk of flesh-like substance he was using to practice in his living room that day. A human-like android with a silicon based substance that looked and felt like skin kept Creedo occupied. Every day he enjoyed practicing being a gladiator. He was getting better at it. The robot could mimic humanlike moves and could react to thrashes and cuts like real people. The robot's sword was a ghostlike hologram, but would make a sharp noise when it would contact. Creedo made a game of this; one contact in the head was worth 3 points, one in the torso--two, and all others--one. Creedo was currently in the lead 8 to 4. The next match was trickier. They both stood their ground. Creedo feigned right, then leap to his left. He raised his shield to block a powerful thrust from the robot, then he pivoted around and stabbed the android in the back. One more point for Creedo. One more for victory. The opponent was far from finished. The metallic creature thrust his holographic lance into Creedo's torso and scored a point for the machine. Creedo only felt a light vibration to simulate contact. The score was now 8 to 5. Creedo was now determined to win. He whipped out his lance at the robot, the robot backed away defensively. He did a pirouette around the robot's left flank. The robot's projection blade clanked against Creedo's weapon. Creedo then ducked down--and in a ballet worthy move--spun around backwards on the ball of his right foot and thrust his blade into the robot's back for the final victory. Creedo enjoyed these games. He enjoyed watching the modern-day games as well. But he yearned for his chance to fight for real. In the mean time this whetted his appetite.

Creedo soon tired of these pretend matches. He thirsted for real blood. It was something primal in every Roman. The need to prove himself as a man. As a Roman. This challenge needed to be done. A worthy opponent was not a mechanical hunk of metal with artificial skin. A worthy opponent was a flesh and blood human who knew it was his life or his enemy's in mortal combat. That's was Creedo needed. Then he would be satisfied--or dead.

Much will change in the future. Much will stay the same. After a succession of presidents and governments; some good--some bad--power keeps centralizing into the hands of the federal government and chief executive. An ever expansive government at home and abroad shreds most individual rights. Power is their only concern. Power is their only God.

Douglas left for Sedona. As he left the motel for the long journey, he put a final message in a plain white envelope. He left the note under the door of an old friend, partner, confidant. The message was long, but it concluded with the words "I forgive you." This letter put an ever slight smile on Stacy Fields' face as he made his way across the country. Douglas made it to Sedona with little trouble. All he had to do now was wait for the signal. But there was still yet unfinished business to take care of. The most important events were yet to come.

The meeting between Creedo and Professor Brown would soon be underway. Similar arrangements between them, and the meeting with myself would be implemented; but now there was at least some established trust among us. This was by far the most important commodity. Nothing more could proceed without trust. Creedo proved that he could be trusted. Which surprised everyone but me.

Burnham blindfolded Creedo again and generally used the same precautions he used when he met with me. Brown was intrigued, but not excited to speak with the senator. Creedo swaggered through the threshold of Burnham's place and darted right to Brown.

"Nice to see you again Dr. Brown. It's been a while. I hope all is forgiven," Creedo blustered.

"Of course. But honestly, I don't know if I can trust you," Brown replied.

"You can I can assure you. We have different opinions, but I hope mutual respect."

"We'll see. So I heard you really wanted to speak to me for some reason. I'd like to know why. It's been years since we last spoke...and had a bit of a disagreement."

"I hope all that's behind us. I actually have a challenge I'd like to invite you to."

"A challenge?! That sounds fun. What could that possibly mean?"

"Or a game of sorts. As you are probably aware, this country of yours has revived the games."

"What games?

"You also call them gladiator fighting."

"Oh that. Yah I saw that on tv. You do know that's fake Creedo. That's all staged reality tv crap for idiots. You didn't actually think that was real did you?"

"I'll have you know that show for idiots, as you say, is for real. And I was the one who first recommended reviving them. It has become quite popular in your nation. I propose we fight in an exhibition match. Since we have difffering philosophies I will represent the American empire; and you represent the American republic. This match will not be to the death like the others. The one who gets knocked down wins. Also, proceeds will be donated to the charity of your choice. What is your answer?"

"Well, I was going to go home tomorrow. But I'd say yes. Let me discuss this with Mike and Jim first. But my answer for now is yes."

"Great. We'll start our training then. I'll show you where the gladiator schools are that offer training. I'm very much older than you, so you have an advantage."

"You look very fit for your age though."

"The advances in medicine are amazing."

"I was afraid of that."

"I just suggest you train well. I'll see you at the games," and with that Creedo exited the apartment and went home. Brown went back to the motel to discuss the agreement with me.

Chapter 40

"I can't believe what you just agreed to!" I exclaimed.

"It's just something he needs to prove. And I'd love to kick the old man's ass!"

"But we need to go home. Didn't you consider that?"

"Look. You wanted to stay this long to get material for your book. After this fight, you'll get plenty more material for your book don't you think."

"I suppose so. So you really want to do this?"

"It's something we have to prove. I can't really explain it as much as feel it. But we're in sync on this. It's something we both need to prove to ourselves."

"Ok. If it's really what you want to do. More power to you. You better train hard for this. I know he's older, but he looks much younger."

"Yah, they have great youth serums here. I suppose it's about an even match. Burnham told me I'd get more logistical information about the fight in the coming days. I'll train in a designated gladiator school, and fight 'till the first one gets knocked down. Then we'll go home."

"Ok. I hope it works out that way. So far nothing's gone according to plan."

"Look, if something goes wrong Douglas will get the order to go home immediately. He can then set things straight."

"Ok, but I don't know how he's going to take this further delay."

"He's used to it."

"We all are at this point."

"Ok. We'll good luck. I suppose I'll watch this fight. Definitely something I'll report back when it's over. I hope you win."

"Either way, Jim goes back to our own time. The outcome is irrelevant."

"Then why are you doing this?"

"I have something to prove too."

Brown was subsequently sent to the gladiator training school built for the games. He was told this was to be an exhibition game where the first one to knock his opponent down would win the game. He would train for six weeks to prepare for the event. Needless to say both Douglas and myself thought he was nuts. To our horror and surprise we had the opportunity to watch a live game in progress. Just like in the days of ancient Rome there was real death involved. We didn't understand how this could happen in such an advanced culture, but Creedo understood better than all of us.

We went to our first game a few days after Brown agreed to the match. We felt like ancient Roman spectators watching gladitorial combat; people being eaten by wild beasts, and live executions. It was horrible; but something one couldn't turn their head away from. One had to watch, like a car crash, or a news story of some horrible event. The justifications were explained to Brown and I thoroughly, but it was still barbaric to watch. Most of the contestants voluntarily chose to participate, and the rest would've been executed anyway; but it was still brutal to watch and amazing that this sport was revived after so many years.

Chapter 41

Brown arrived at his scheduled training class early. His teacher was a drill-sergeant type with a crew cut, six-foot frame and a muscular physique. Brown's first days were hell. Mostly routine but harsh exercises, he needed to be knocked around a lot before he was built up to fight. Burnham and I checked on his progress, and to give him moral support. He was a quick learner, after a few weeks, he learned the techniques of the ancient weapons. Still, the betting on the future match was heavily in favor of Creedo. After extensive training, we saw Brown's first practice match. He practiced with wooden swords and pillow shields. His opponent was a middle-aged man in reasonably good shape. Weighing about 20 pounds more than Brown, he had a slight weight and reach advantage. Brown, however, did much better than expected. They entered a ring raised in the center of a large gymnasium. Brown's opponent lunged into him thrusting his bamboo sword at his chest. Brown quickly deflected his thrust and counterattacked. Brown struck on the upper part of the man's back. That was enough for a point in their scoring system of a ten point match. His opponent quickly came back. A quick jab to the head for the opponent gave him the next point. The next points all belonged to the opponent. Brown lost, but his tenacity and perseverence impressed his trainer and his opponent. That was more important than anything else. When the match was over Brown approached me outside the ring.
"You did great Brown!" I exclaimed.
"Thanks, not bad for a beginner."
"Not bad for anyone. I'm not an expert or anything, but you seem to be a natural."
"I guess. I still need a lot a practice."
"Keep at it my friend. Keep at it."
"I'll beat that Creedo...you'll see. I'll beat him."
"If you do, you'll beat the odds. After this is done what a story we'll take back with us. What a future this is!"
"Yah, it's not what I expected. No one could've predicted this. Well Mike I gotta finish training."
"Ok. Good luck, see you soon."
"You too."

Creedo was in his private training room practicing his gladitorial skills with the best trainer in the country. Already somewhat of an expert fighter with the gladius, he needed to refine his moves for the upcoming match. Practicing with real armor and weapons, he was upgrading his fighting capacity and ready for the real thing. He built himself up with the help of his new drugs and exercise program. As fit as any thirty year old, Creedo was ready for this match and challenge. What Brown didn't realize was Creedo was ready and willing for anything and to do anything.

Chapter 42

Douglas called Brown on his cell. The conversation was short, but to the point. He was tired and wanted to go home. He couldn't understand why Brown would prolong this trip just to prove his manhood with a silly game.

"Hey Brown, what are you doing man. Let's just go home!"

"We will. Just be patient, this will be over soon."

"I don't want to wait any longer. What do you have to prove anyway?"

"It's about more than me or Creedo. This may sound silly but I think the future of our country is at stake over this match."

"What do you mean? The future's already been set. One match has nothing to do with the anything."

"Look, it's just something I have to do. When you guys get back to our own time you can tell them that one man risked his life for the future of the republic."

"What do you mean risk your life? I thought this was an exhibition."

"Jim...I believe this is a fight for our lives."

"I don't know what you've gotten into Harry. I hope you know what you're doing."

"I'm learning. I'm being trained well. I'll beat him, trust me on this one."

"Whatever happens, Mike and I will go back and make things right. I'll watch the match on my computer. Good luck."

"Thanks...you too Jim."

"Bye."

Brown's practice sessions were getting more intense. He learned with real swords and in some training matches got hurt pretty bad. He insisted on continuing with the match however; it seemed like the harder the training was on him--the more he wanted to fight. It's a tenacity I hadn't seen in him before, and hadn't seen in hardly anyone else I knew...except Creedo.

Back at the hotel I was getting anxious. I wondered why Brown agreed to this. We could've just gone home and made this all so much simpler. But he had something to prove. Douglas and I were getting pissed; but we could understand. Vicariously, we had something to prove too. And through Brown's match this was our fight as well. I began to speak to and get a sense of the people in this future world. The government was powerful and omiscient. But the people weren't any happier or more prosperous because of it. Powerful empires only help the elites, and this society was proof of that. Far off imperialist wars abroad didn't help the people at home. But they were clearly diverted by the entertainment of the day. My friend Brown would be living proof of that. In the future all the politicians were pigs at the trough of the military-industrial complex. There were no Lou Parns left anymore.

In my journal I protested Brown and Creedo's stubborness. I jotted down some observations about the country. I watched some television--heavily promoting the upcoming match, then fell asleep. I was looking forward to the end of this.

Chapter 43

In order for Brown to be incognito as far as the authorities were concerned; he needed a new identity. He also needed a backstory. This Creedo and his lackeys gladly provided. His new name was to be Ed Hayes. He underwent some extensive prosthetics on his face to alter his appearance. His story was that his wife had a terminal disease and he needed the potential reward money from winning the match to pay for surgery. It was a story that was both sympathetic and inspirational. Of course, his association with me or Douglas had to be expunged as well. This all had to be completed well before the match began. During this time period, Creedo asked to speak to me in private. He sent for a private limo to pick me up and take me to his penthouse for a little brunch and chat. When I arrived he opened the door and took me to his patio.

"Hello Mike. I have some questions for you before the match begins," he opened.

"Good. I've been wanting to talk to you too about this! Nice to speak to you in private finally," I replied.

"Do you trust me?"

"I think that's a question for Harry. He's the one you're going to fight. As for me, let's just say I play it safe. We have many back-up plans if this fight doesn't turn out well for him. The second anything goes wrong; well lets just say we're prepared."

"Good. You were probably surprised he agreed to this."

"That goes without saying."

"I know the reason he agreed to this. It's the reason we both agreed to this match."

"What is it Creedo?"

"Destiny. This match was destined to happen. Like Father Aeneas and Turnus is Virgil's great poem, this fight is for the destiny of two futures. He knows this and so do I. It is a literally a fight for the future of your country."

"You know I really don't care about that so much. I just care about my friend. If anything happens to him; well like I said we have a back-up plan that can change anything."

"I know that Mike. And I'm fine with that. Still the importance of this match cannot be underestimated. It is for the future of your republic. I just wanted you to know that."

"The main thing to me is the future of my friend. I just want you to know that. Honestly I didn't want him to fight, but like you he's very stubborn." Looking around at his opulent furnishings I stated, "you've done very well for yourself here."

"Yes I have. I helped inspire the path your country has been on for the last 30 years. I have acquired many important positions with your government. Thanks to you and you taking me here my life has been happier and more fulfilling than it could have been in Rome. Now I must crown my achievements with this victory in the arena."

"If that's what you want. After the match is over, we're going home. To our time. We won't interfere with your success, or the country's path or anything. I just want to take my friends home safe and sound. That's my only goal."

"Then may the gods be with you my friend. May the gods be with you."

With that Creedo bowed ever so slightly towards me. Despite my better judgement, I really felt at this moment he could be trusted. He escorted me to the threshold and I walked confidently back to the limo and my motel.

Chapter 44

"So I guess I'll be calling you Ed from now on?" I teased the professor while in his training room.

"You can call me whatever you like when we're alone. But if there's anyone else present then call me Ed. Don't blow my cover. I really believe I've really improved at this. I'm in much better shape and my fighting skills have gotten much better."

"Good. So I know who to bet on," I said sarcastically. "Just stay safe Ed, Brown, or whoever you are," I then left to let him train.

Douglas was holed up in Sedona at a fleabag motel waiting for the word. He was about to investigate the facility where the time displacement unit was stationed; where the government didn't know about it. I sent him the coordinants by GPS to his cell. He immediately disposed of his cell so it couldn't be traced. Located underneath an abandoned warehouse, Douglas had to see if was still there. It was; but he still had to test it. He had to see if it worked by going 10 minutes forwards or backwards in time. He decided to go forward 10 minutes to test it. When Douglas stepped out of the cylinder-like structure he walked out of the warehouse. He saw two men approaching. Both had buzzcuts, dark shades, suits and shoes. He knew who they were and knew how to react. They briskly came towards him.

"Agent Douglas, NSA," he proclaimed.

"FBI : Agents Lord and Fisk, can I see some ID," they demanded.

"Of course," Douglas showed his badge and identification. "We'll handle this."

"We were ordered to guard this area. What's NSA doing here," Lord responded.

"It's classifed. But you guys are relieved of duty here. I'll stand watch here. I wish I could tell you why, but you know I can't."

"I hate this bureaucratic bullshit. They never tell us anything! Just because we only been with the feds for a year doesn't mean they should treat us like shit! Well have fun guarding a warehouse Douglas. I'm not blaming you Douglas, but this isn't the first time we got the runaround like this," Fisk ranted.

"I'm sure its not. Man if I had time the stories I could tell you. We're all just pawns here. You've just started getting shit on, believe me."

"Well hey good luck Douglas. We'll see you." With that the agents left, and Douglas returned to the warehouse. He knew this area wouldn't be safe for very much longer.

Douglas called me from his cell. Clearly pensive, he spoke quickly.

"Hey Mike...you better make this quick. They're on our tail."

"What happened Jim?"

"Two agents in Sedona. I diverted them for a while, but this fight better be quick!"

"Okay. Well it's only a week away now. I hope that's enough time."

"Well, we can have enough time...we have access to a time machine remember."

"Right, we'll put that to use if we have to."

"We may have to."

"Ok Jim. Keep me posted."

"I will Mike." He then hung up. I now wanted this match over with sooner than ever.

Chapter 45

Creedo arrived at my doorstep to speak to me. He wanted to talk to me one last time before the match. Looking confident and much younger than his years, he obviously had much to reveal about the upcoming fight.

"This is probably our last meeting together, I just want to say thank you for being a friend, for bringing me to this world, and showing me everything it has to offer. You will be in my heart forever Mike."

"Thank you. You have adapted better than I could have ever hoped. I offered to take you back to your own place and time, but you preferred to stay here. I was surprised, but respect your choice."

"This is my time now. This is my destiny. You have delivered me to my destiny. After this match I will meet my destiny."

"What do you mean? This is an exhibition match. No one is supposed to get hurt. You'll be okay."

"This match is my destiny. And your friend knows this too. Nothing can change our destiny. You need to realize this."

"Honestly, I've been against this match from the beginning. But unfortunately it's not up to me. I just can't wait to go home."

"May the gods be with you my friend." With that he gave me the traditional Roman salute and walked away. After that I wanted to talk to Brown one last time before the match. I wanted to see if he shared the same fatalistic attitudes as Creedo.

Brown was doing one armed push-ups in the large gymnasium when I arrived. He was much bulkier and muscular than when I last saw him. He told me he was taking the same youth-drugs Creedo was taking to help even the match. I responded that I thought that was fair; even though Creedo was older, he seemed more fit than did Brown, so any enhancement for Brown would be acceptable. The rules of the match would be as follows: They would use authentic weapons carried in a belt-pouch around their midsection; they would wear impenetrable armor made from a steel-titanium alloy so they wouldn't get hurt; the first person to knock down their opponent on the ground would be declared the winner. The weapons they would use would be a Gladius- a short sword; a Sica- a double-edged curved sword 16-18 inches long; a Scissor- a steel tube that encased one's forearm with one's hand end cut off with a semi-circular blade attached; a trident and a net- to throw over one's rival or entangle their weapon; a scutum- a rectangular shield; and a cestus- a special protective glove. Both Creedo and Brown were anxious and excited for the match to begin. Douglas and I just wanted to go home.

Chapter 46

The day of the match was finally at hand. This was a very big day. It started off with lions and tigers devouring condemned prisoners just like in ancient times. Then they proceeded with the voluntary gladiatorial combat for prizes. I reluctantly witnessed this spectacle watching in horror as real people were at first eaten, then risked their very lives for entertainment. Brown and Creedo's match would be last. My seat was in the fifth section back from the middle of the amphitheater's oval. I got a good view and with the giant video screen projecting every gruesome detail, I got a visceral experience from what went on in the arena. The first round of "entertainment" for the day consisted of condemned prisoners who chose to be devoured by lions, tigers and other wild animals. They chose this method of execution so their family and loved ones would be amply compensated. In this future, this is considered the most noble form of execution for that reason. The first man condemned in this fashion was dragged in shackles to the center of the arena. He then had blood splattered on him and a meat-like odor sprayed on him to ensure the animals would attack. The man stood there as still as a tree. Then, a ravenous lion materialized like out of nowhere. Of course, there was an extensive sublevel inspired by the hypogeum of ancient Roman amphitheatres where animals could appear almost anywhere in the arena. The lion lunged at the poor man, bit his throat while the crowd roared its approval. I personally couldn't bear to watch, but I learned what happened and all the details by a very pleased spectator. The next act was a very standardized pop band that gave a performance. The type of act that typically performs at football games and other events in our own time. It's very similar in the future; except death is the game, instead of touchdowns, hits or runs.

After much fanfare the final match began. The first combatant to come out was Creedo. A large iron gate opened on the right side of the arena. Creedo strided out confidently and the crowd roared. His weapon belt was full to the hilt. His steely eyes screamed he was ready for battle. The announcer introduced him to the public. They liked the fact he was an ancient roman citizin participating in the new games. To the crowd he brought an authenticity to the game that the other contests lacked. After Creedo's introduction was done, he stood motionless at the center of the arena floor. From the left side of the arena, a gate opened and Brown came through. He, too, confidently strided to the center of the arena. They then politely acknowleged each other, shook hands, and waited for the signal to begin. The signal came when the announcer shot a starter pistol. Creedo attacked first; thrusting his Sica downward at Brown. Brown quickly moved away, pulled out his gladius and glanced Creedo's breastplate armor with it. Creedo deftly swung around and struck Brown's armor on the back giving him one point out of the three for the match. Brown was clearly irritated, so he ducked down, flanked Creedo's left side and threw a net at his eyes. Creedo then was distracted long enough for Brown to tap Creedo with his short sword on his chest, tying the match at one point. The crowd cheered wildly at this. But Creedo didn't pause for a moment, he deftly flanked Brown's left side, drew out his sword--motioned that he'd swing left, but swung to his right side and tapped him on his right arm for the second point. Brown pulled out his long sword. After about five minutes of blocking Creedo's thrashes, Brown made a lucky shot at Creedo's chest area. To his great surprise it made contact, tying the score at two. The next point would win the match. They started out fighting more aggressively. They both swung and blocked frequently. However, Creedo seemed to be fighting more aggressively and getting closer to contact with Brown's armor. The fight went on for over ten minutes. Finally, Creedo locked a frozen glaze at Brown. A spirit rushed in him like I never seen before. Creedo swung wildly attacking brown until it struck Brown's helmet denting it slightly. Brown fell back abit. Shook himself up and stood up. By then it was too late for Brown, Creedo got the final point and won the match.

The match was seemingly over then. The crowd roared and the announcer congratulated Creedo on his win. Creedo then shocked the crowd by taking off his armor and saying something to Brown. I later learned was transpired between them.

"I'll give you another chance to win the match. If you're prepared for the ultimate challenge," Creedo exclaimed.

"I'm prepared as much as you Creedo!" Brown replied. With that Creedo took off his armor and signaled Brown to do the same. Brown responded to my shock and horror, in agreement, and threw off his protection as well.

"Now the people will witness a real match." Creedo shouted.

"I'm as ready as you are! We'll end this once and for all!" Brown retorted. With that the real match began. Creedo swung first. Brown swiftly averted the blade by leaning backwards. Creedo shouted something at Brown. He said something about the republic being over and no match would change that. Brown retorted that may be true, but Creedo would be defeated and the republic would be restored.

"Your republic is dead, and you shall be too!" Creedo replied.

"I will use every last breath of energy to defeat you!" Brown then took an enormous swing at Creedo's gut. Creedo was grazed slightly, but was relatively uninjured. This fueled Creedo's agression, he responded stronger than ever. Creedo could feel Brown's hesitation in his next move. The crowd could feel the violence between them. They were clashing blades back and forth for what seemed like an eternity. Then suddenly Creedo broke out of the rut. He swung his blade at an unprotected spot on Brown's neck. With the whole arena going silent, Brown fell to the ground for the last time. Creedo's face appeared solemn. He looked up at the audience for approval. They remained silent. Then it erupted in roars and shouts. The winner of the match and the future was Creedo. He bowed in all directions and slowly walked out of the arena victorious.

Chapter 47

Shocked and horrified that Brown would participate in such a spectacle, and lose his life over it, I strolled with trepidation out of the arena. Determined to go back to my own time to warn people of their upcoming future and to save my friends from such a fate, I was surprised to be accosted by a man claiming to be a friend of Creedo's. He told me to meet him at his penthouse before going home. Just to be on the safe side; I ordered Douglas to take off if anything happened to me. After this I would go back to the present and restore everything to how it should be I kept telling myself, but first I had to do this. So, almost possessed by some unnamed force I went. When I arrived at Creedo's door I knocked with reluctance. He happily stated to come in while he was sitting at his mahogany desk spinning a large globe with large swaths of blue representing American power and influence all over the world.

"So what do we have here Creedo?"

"An empire...if you can keep it."

THE END